COLOUR:
A PHILOSOPHICAL INTRODUCTION

Aristotelian Society Series

Volume 1
COLIN McGINN
Wittgenstein on Meaning:
An Interpretation and Evaluation

Volume 2
BARRY TAYLOR
Modes of Occurrence:
Verbs, Adverbs and Events

Volume 3
KIT FINE
Reasoning with Arbitrary Objects

Volume 4
CHRISTOPHER PEACOCKE
Thoughts:
An Essay on Content

Volume 5
DAVID E. COOPER
Metaphor

Volume 6
DAVID WIGGINS
Needs, Values, Truth:
Essays in the Philosophy of Value

Volume 7
JONATHAN WESTPHAL
Colour:
A Philosophical Introduction

Volume 8
ANTHONY SAVILE
Aesthetic Reconstructions:
The Seminal Writings of Lessing, Kant and Schiller

Volume 9
GRAEME FORBES
Languages of Possibility:
An Essay in Philosophical Logic

Volume 10
JONATHAN LOWE
Kinds of Being:
A Study of Individuation, Identity and the Logic of Sortal Terms

This volume edited for the Aristotelian Society by Anthony Savile and David Galloway.

Jonathan Westphal

Colour: A Philosophical Introduction

Aristotelian Society Series

Volume 7

SECOND EDITION

Basil Blackwell

First published (as *Colour: Some Philosophical Problems from Wittgenstein*) 1987
in cooperation with The Aristotelian Society
Birkbeck College, Malet Street, London WC1E 7HX

New paperback edition 1991

Basil Blackwell Ltd
108 Cowley Road, Oxford OX4 1JF, UK

Basil Blackwell Inc.
3 Cambridge Center
Cambridge, Massachusetts 02142, USA

British Library Cataloguing in Publication Data

A CIP catalogue record for this book is available
from the British Library.

Library of Congress Cataloging in Publication Data
Westphal, Jonathan, 1951–
Colour: a philosophical introduction/Jonathan Westphal.—
New pbk. ed.
p. cm.—(Aristotelian Society series; v. 7)
Previously published as: Colour: some philosophical problems from Wittgenstein. 1987.
Includes bibliographical references and index.
ISBN 0–631–17934–8
1. Color (Philosophy) 2. Color vision. 3. Wittgenstein, Ludwig, 1889–1951. Bemerkungen über die Farben. I. Title. II. Title: Color. III. Series.
B105.C455W47 1991
121′.3 — dc20 90-43858
CIP

Set in 11/12pt Times by Graphicraft Typesetters Ltd., Hong Kong
Photoset, printed and bound in Great Britain by
Billing & Sons Ltd, Worcester

Contents

To Steve Graham

Acknowledgements

A version of Chapter 2 of this work appeared in *Mind* for July of 1986, along with some paragraphs from Chapter 6. The same material, with an Introduction on Wittgenstein and Goethe, formed an article in *Goethe and the Sciences: A Reappraisal*, edited by Fred Amrine, Francis Zucker and Harvey Wheeler, in the Boston Studies in the Philosophy of Science, 97 (Dordrecht: Reidel, 1987). Chapter 2 also incorporates 'White in All Possible Worlds' from the *Proceedings of the 7th International Wittgenstein Symposium* (Vienna: Hölder-Pichler-Tempsky, 1983). Chapter 3 is a revised version of 'Brown', *Inquiry*, 4 (1982). 'Black', Chapter 5 of this new edition of *Colour*, is from *Mind* of 1990. The new Chapter 8, 'Simplicity', is a revised version of 'The Complexity of Quality', *Philosophy*, 59 (1984).

I have benefited from discussions with David Wiggins, Jerry Lettvin, Elizabeth Hackett, Hilary Putnam, Elisabeth Anscombe, Natalie Adams, Michael Wilson, Justin Broackes, Mike Land, Michael Slote, W.D. Wright, the Epiphany Philosophers, Larry Hardin, Bill King, Hans Kamp, Anthony Savile, Kit Fine, Mark Sainsbury, Carl Levenson, Christopher Peacocke, and from the suggestions of two referees. I am grateful to all of these people, in particular because I have borrowed a number of points in the text from them. Ralph Brocklebank read and commented on the entire work in manuscript. Carol Scott of Idaho State University provided valuable secretarial help.

J.W.

Farben regen zum Philosophieren an. Vielleicht erklärt das die Leidenschaft Goethe's für die Farbenlehre.

Die Farben scheinen uns ein Rätsel aufzugeben, ein Rätsel, das uns anregt – nicht aufregt.

WITTGENSTEIN

My love of nature consisted mainly of my enjoyment and my enthusiasm for colour. Frequently I was so obsessed by a fragrant and strongly resonant blue spot in the shadow of a shrub that I would paint a whole landscape only in order to pin it down.

KANDINSKY

1

Introduction

> In philosophy we must always ask: 'How must we look at this problem in order for it to become solvable?'
>
> WITTGENSTEIN

1.1 The Puzzles

In *Remarks on Colour*[1] Wittgenstein discusses a number of puzzling propositions about colours. This monograph is about the explication of these propositions, which I shall call the 'puzzle propositions'. Representative examples are:

(i) Something can be transparent green or any other colour, but not transparent white;
(ii) White is the lightest colour;
(iii) Grey cannot be luminous;
(iv) There cannot be a pure brown or brown light;
(v) There is no blackish yellow;
(vi) There can be a bluish green but not a reddish green.[2]

The puzzle propositions express striking and initially surprising facts about the colours, and all of these facts seem to be of the same kind. A question to which I give an answer in this monograph is whether the impression that there is one such kind is correct, and, if so, what it is. This amounts to one sort of answer to the question 'What is colour?'

Corresponding to each of the puzzle propositions is a puzzle question. Why can't something be transparent and white? Why is white the lightest colour? Why can't grey be

[1] Ludwig Wittgenstein, *Remarks on Colour*, ed. G.E.M. Anscombe, trans. Linda L. McAlister and Margarete Schättle (Oxford: Blackwell, 1978).
[2] Ibid., (i) I 19, (ii) III 1–3, (iii) I 36, (iv) III 60, 65, 215, (v) III 106, (vi) I 9–14.

luminous? Why can't there be brown light or a pure brown? Why is there no such thing as a blackish yellow? What is wrong with the very idea of a reddish green? I shall offer answers to the puzzle questions which are all of exactly the same kind. The basic principle is that colours are a certain sort of shadow.[3] What this means will become clearer as we proceed. One crucial interpretation is that for a material or substance to be coloured a colour is for it to exhibit a particular phenomenalistically interpreted absorption spectrum.

1.2 A Method for Solving the Puzzles

The method I shall use to answer the puzzle questions is as follows. Each of the colours and the properties involved in the puzzle questions, such as being transparent, receives a definition. I shall express this by saying that the colours and the other relevant properties are 'analysed'. The definitions which provide the analyses are real definitions in the sense that they are definitions of the colours and the other properties, and not of the words which are their names. So by a real definition I shall mean a definition which states what the colour (or being coloured the colour) is, and in the sense in which I am using it it could be called a theory, or perhaps a *theorita*, a little theory, to appropriate Richard Braithwaite's charming diminutive term. At any rate it is a part of a scientific theory. The puzzle propositions are then deduced from the definitions. So the conjunction of the negation of the puzzle propositions and the definitions is a contradiction. Sometimes I shall demonstrate the deductive relationship between the puzzle propositions and the real definitions by giving the deductions, and sometimes by generating the contradiction between the definitions and the negations of the puzzle propositions. The difference is one of convenience, and the choice depends on the form

[3] This is Goethe's *Grundgedanke*. Cf. *Remarks on Colour*, III 57. The best general introduction to Goethe's science is Erich Heller, 'Goethe and the Idea of Scientific Truth', in *The Disinherited Mind* (London: Bowes, 1975), and also J.W. von Goethe, *Scientific Studies*, ed. Douglas Miller (New York: Suhrkamp, 1988).

in which the different puzzle propositions are stated by Wittgenstein.

These formulations are not uniform. Sometimes Wittgenstein expresses the puzzle propositions in modal terms, as in (i), and sometimes not, as in (ii). Sometimes he offers little more than a fragment of a sentence to suggest a puzzle question. I imagine that it would be easy enough to give a uniform formulation to all the puzzle propositions, either modal or non-modal, and also to convert all the puzzle questions into the corresponding puzzle propositions or vice versa. The danger to this procedure would be that it ignores the fact that the explication of the logical status of the puzzle propositions is a very delicate one, and much depends on capturing the exact sense which Wittgenstein intended them to convey. I have chosen to leave the text of *Remarks on Colour* as it is.

Applying the method described above shows why the puzzle propositions are necessarily true if they are true. Given the real definitions or knowledge of what colours are, the denial of the puzzle propositions is a contradiction. A contradiction results when the real definitions are substituted for the colour terms in the denial of the puzzle propositions. But of course for this to be an explication of the necessity of the puzzle propositions, the definitions must themselves be necessarily true. I subscribe to the view that it is necessary that everything is what it is, both in the sense that my cat could not have been some other cat, though I might have been the owner not of this cat but of some other, and in the sense that everything is necessarily the *kind* of thing that it is. *Pace* Kafka and Ovid, I think that it is necessary that I am a man, or a woman – a human being – and not a bus shelter, or a beetle, or anything else. But this is not my topic. I shall assume the necessary metaphysical principle, and this monograph can be regarded as tracing the consequences of the conjunction of this principle with a certain kind of phenomenalism. My answers to the puzzle questions can then be construed as arguments for the existence of essences of colours satisfying the phenomenalistic real definitions. We could of course put off the day when the necessity of the puzzle propositions was to be explained, either by a principle which was itself necessary or by one which was not. What I have shown is not that the transition from the modal to the non-modal must take place within the definitions, but that the propositions which generate the puzzle propositions, whether their truth is in the mode of necessity or not, have the power to integrate *all* the facts collected by the puzzle propositions. No essentialist could want more.

So the task I have set myself is not the explication of the modality of the puzzle propositions, nor of the proposition that colours are necessarily whatever they are. My purpose is merely to answer the prior question what colours are, and to show how the answer which I give is derived from the puzzle propositions. The puzzle propositions provide a kind of test which any successful theory of colour must pass. They are puzzles about the colours we see, are directly aware of, or whatever, and what explains the puzzle propositions is the 'what it is to be' the relevant colours. The puzzle propositions therefore also provide a very strong set of constraints on any programme for reducing colour theory. Indeed, the puzzle propositions provide a means for *determining* what colours are. We can work backwards from the puzzle propositions to the definitions of colours. And the *necessity* of the puzzle propositions shows that with them we are very near to the essences of colours.

1.3 Topics Avoided

I have tried as far as possible to avoid the topic of colour identity and sameness of colour. My defence of this omission is that discussion of the relevant points would take us too far into the general philosophical topic of identity. There are of course special questions concerning identity which arise in connection with colour and colour science, such as the relation of ordinary judgements of sameness and difference of colour to the concepts of colorimetry and colour measurement, the precise formulation of identity statements in connection with contrast and analogous hue effects, 'real colour', the alleged breakdown of transitivity for colours and for shades of colours, the analysis of 'This is the same colour as that, but they are not the same shade',[4] and so on. I have also excluded anything to do with the more tedious and unrewarding philosophical colour topics, such as spectrum inversion, other minds, new colours in other possible worlds, etc.

The discussion is cut off at various points for reasons of space, and I shall say little or nothing about a number of important philosophical topics raised by Wittgenstein, including for example the thesis of the indeterminacy of sameness of colour and the suggestive point he makes at *Remarks on Colour* I 32 about the duck rabbit-like effect of the shift of the puzzle propositions from the empirical to the logical mode. And readers bracing themselves for yet another Necker cube can relax. I do not discuss colour or any other illusions. My main concern is with the clearly logical puzzles about colour raised by Wittgenstein, and I have avoided attendant questions in the philosophy of language, metaphysics and epistemology, except where they contribute more or less directly to the answering of the puzzle propositions. Chapter 9 is about the implications of the answers to the puzzle questions for the philosophy of mind. There are also in *Remarks on Colour* puzzle propositions relating to colour effects involving spatial and temporal factors, such as scintillation and glitter, which I shall not consider. *Remarks on Colour* consists of a number of very delicate and powerful arguments, and the present work is a commentary on the structure of just one of these arguments.

I have also tried to steer around questions concerning the basis and justification of the linguistic cuts in colour space. It might be thought that this would make it difficult, if not impossible, either to say what colours are, or, by that means, to solve the puzzle questions. This turns out not to be the case. What does emerge is the distinction between colour considered as a category imposed on or implicit in colour space, and colour considered as so to speak a real phenomenon, almost a sort of stuff. This distinction is very

[4] See my 'Universals and Creativity', *Philosophy*, forthcoming.

hard to pin down and make precise, but the type of ambiguity I have in mind was meticulously sorted out by Aristotle for number at *Physics* 219b 6–9: 'Number, we must note, is used in two senses – both that which is numbered and the object of enumeration we call a number, and that by which we enumerate'; and for colour in *De Anima* II 6, in connection with 'special sensibles' such as colour, 'about which we cannot be deceived', and which cannot be perceived by another sense, and 'incidental sensibles' such as the son of Diares, who is perceived incidentally when we perceive a certain white body – his. Aristotle's distinction between the special sensible, which is the colour, and the incidental sensible, which is coloured a colour, becomes crucial in Chapter 7, in which 'X is reddish green' and 'X is coloured red and green all over at the same time' are distinguished. The distinction between colour considered as a real phenomenon of the senses and a constituent of the real world, and colour considered as a category or abstract division of colour space, however unclear it may be, does explain one otherwise puzzling feature of the philosophical literature on colour. What I have in mind is the existence of a quantity of excellent philosophical work on questions about colour and colour concepts which connects hardly at all with the scientific or artistic study of colour. Bernard Harrison's *Form and Content*[5] would be an example. Very roughly, Harrison's book addresses the question, 'What are colours?', meaning by this a question about colour concepts or categories, i.e., what we do with words; whereas, anthropology and linguistics apart, colour science is concerned with the different question, 'What is colour?', the dimension in which colours are scaled. I do not myself see how the gap between work based on the two different senses of 'colour' could possibly be unbridgeable. Yet it would require the concinnity of two concepts of a concept of a colour yielded by a distinction parallel to Aristotle's: (i) a concept is a use of a word, a saying (in the sense of an uttering, rather than an utterance or thing uttered); (ii) a concept is what is provided by a *theorita*, an objective principle in

[5] Bernard Harrison, *Form and Content* (Oxford: Blackwell, 1973).

virtue of which a thing is what it is, to be distinguished from the conception or psychological particular involved in the production of (i), if there is one.

There is an obvious and Wittgensteinian objection to this method. The limits of the application of a colour term such as 'white' must be settled *before* any question of the relation of *that* colour to a scientifically conceived real essence can be raised. So we must know what counts as the colour white, or what the colour white is, *before* we can know anything about the physics of it. This is really an argument for linguistic idealism, the thesis that essence is grammar. I think the point can be conceded. What it means is that there is or must be a preliminary classification or phenomenology given by the language. Two levels of analysis can be distinguished, corresponding to two distinct senses which can be given to the question 'What is it?' asked of a colour. (1) In the first sense, an appropriate answer might be 'red'. Here 'What is it?' means '*Which* is it?' (2) In the second sense, 'What is it?' means 'What is the thing which is meant?' (1) certainly is the prior question.

1.4 The Significance of the Puzzle Propositions

The importance of propositions similar to the puzzle propositions in the history of twentieth century philosophy is well known. Colour incompatibility destroyed two central theses of the *Tractatus*: extensionality, or the truth-function theory of complex propositions, and the independence of elementary propositions. Wittgenstein's failure to fulfil the implicit construal of *Tractatus* 6.3751 (colour incompatibility is logical impossibility because the 'logical structure' of colour rules it out), and his belief that finally no analysis of the sort he had anticipated could fulfil it, were responsible for the advance from the one true logic of the *Tractatus* to the self-generating logics or language games of the later Wittgenstein. This belief caused Wittgenstein to extend his notion of what logic must be, and to alter his idea of the nature of a philosophical problem.[6] Thus he could write in the *Remarks on Colour* that,

[6] See Anthony Kenny, *Wittgenstein* (London: Penguin, 1973) pp. 103–119.

> It is not the same thing to say: the impression of white or grey comes about under such and such conditions (causally), and to say that it is the impression of a certain context (definition). (The first is Gestalt psychology, the second logic.)[7]

There is of course a difference between the two things which Wittgenstein says it is not the same thing to say. I have tried to show in the present work that all the same logic and psychology do meet in metaphysics. What comes about is the coming about of *something*, and what is, here an impression in a certain context, is, so described, what it is because of the way in which it has come about.

Questions similar to the puzzle questions also attained a special importance in the wake of logical empiricism. The puzzle questions are not either clearly analytic or clearly synthetic. Today this is of less immediate concern to us, because we are much more suspicious of this type of distinction. If the chosen examples do not fit the classification, all that this suggests to us is that we should scrap the classification. The examples have in themselves nothing problematic. They only appear so in the distorting light of the distinction, and its superposition on the quite different distinctions between a priori and empirical concepts and knowledge, and necessary and contingent truths. Indeed, the various claims made about the classification of puzzle type propositions were themselves the philosophical problem rather than the answer to anything. The genius of Wittgenstein is an exception. In the *Remarks on Colour* he introduced new and imaginative examples, including (i)–(vi). The purpose of his discussion of these examples was not merely the organization of ultimately unworkable distinctions.

The answering of the puzzle questions is however of more than a purely historical interest. I have tried in what follows

(1) to establish that real definitions of the colours (*theoritas*) can be given, by

[7] *Remarks on Colour*, III 229, cf. I 51.

(2) giving the real definitions of the colours;

(3) to explain by means of (2) why the different colours have the asymmetrical properties they do (e.g. the impurity of brown), i.e. to deduce the puzzle propositions from the real definitions;

(4) to suggest the basis of the truth of a certain sort of relational common sense realism about the being coloured of coloured objects and materials, based on (2);

(5) to show, on the basis of (1), that no special 'logic of colour concepts' is needed to answer the puzzle questions, as Wittgenstein came to believe it was after 1929, and that the analyses given under (1) show that the impossibility of (i)–(vi) and the other puzzle propositions is a *logical* impossibility in the sense of the *Tractatus* after all;

(6) to show by (1)–(3) that colours do have essences, construed as whatever it is that is articulated by a real definition;

(7) to highlight the fact that the definitions which answer the puzzle questions are neither physicalist nor mentalist, and to establish a type of phenomenal description for colours which is neutral with respect to the mental and the physical, or to the phenomenal, conceived as it is in some current literature as a species of the mental. The real definitions can be given either a physical or a phenomenal interpretation;

(8) to show that colours are not simples. This follows from (2);

(9) to show that there is no conflict between science, correctly interpreted, and common sense.

In Chapter 6 I attack physicalism, but 'mentalism', if that is the right name for the view I have in mind, has not been stated in a way precise enough to offer any prospect of answering anything as definite as the puzzle questions. Maybe there is something to the idea that the puzzle propositions are 'phenomenological laws' of our subjectivity 'governing how things can seem, these laws arising from the very nature of perceptual experience.'[8] If so, it must be shown how they can be deduced from a proposition or propositions stating the nature of perceptual experience. The argument of this book is that the 'phenomenological laws' given in (i)–(vi) are the direct expression, not of the nature of our perceptual experience, but of the nature of colour, the nature of the different colours, and what it is for something to be coloured.[9]

The importance of (7) lies partly in the interpretation and assessment of those of the *Remarks on Colour* which refer, directly or indirectly, to Goethe. The reader should be warned that my view is that the relevant parts of the *Theory of Colours* are straight science, not poetry or phenomenology, in the philosophical sense, and not, as Wittgenstein believed, a logical inquiry mistaken by Goethe himself for a scientific one. So I am attacking the suggestion made by Wittgenstein at *Remarks on Colour*, II 16, that

> Phenomenological analysis (as e.g. Goethe would have it) is analysis of concepts and can neither agree with nor contradict physics.

In evidence there is the crucial claim made by Goethe, which is at the centre of his polemic against Newton, that (as we would say) colour is an edge-phenomenon. Wittgenstein wrote:

> This much I can understand: that a physical theory (such as Newton's) cannot solve the problems that motivated Goethe, even if he himself didn't solve them either.[10]

[8] Colin McGinn, *The Subjective View* (Oxford: Oxford University Press, 1983) p. 37.

[9] McGinn also writes (ibid.) that the claim that there are phenomenological laws governing how things seem, construed as 'laws of subjectivity', which may be derived from the nature of perceptual experience, is simply the conjunction of the claim that colours are secondary qualities and the existence of internal relations between colours. If this is so, then, since it can be shown that (i)–(vi) are not 'laws of subjectivity' because they are direct consequences of the definitions of physical colours, it follows that colours are not secondary qualities.

[10] *Remarks on Colour*, III 206.

Under (7) I am in effect claiming that it is not the fact that Newton's theory is a *physical* theory that raises the question whether it has the competence to handle Goethe's problems. The difficulty is rather that the physical theory is unclear in its foundations and its relation to the facts is problematic. There is a serious question, for example, whether and in what sense white light is either white or light. Corpuscles or electromagnetic waves are not themselves light or dark. It will not do simply to consign the problematic phenomena to subjectivity and leave the rest to physics. In Chapter 2, I show how the core of Goethe's theory concerning the nature of whiteness is a physical theory, which is designed to take the phenomenalist interpretation which I give it. I hope to make the reader feel how the kind of semi-philosophical gibberish encouraged by the *Opticks* could have so aroused Goethe's venom. How, to take a later example, could Maxwell's ear have betrayed him so badly that he could write this:

> It seems almost a truism to say that colour is a sensation; and yet Young, by honestly recognizing this elementary truth, established the first consistent theory of colour.[11]

It would be nice to know what Maxwell thought he meant by that first logically ungrammatical clause, and it would also be instructive to establish the role which Thomas Young's understanding of the 'truism', whatever he thought it meant, played in the development of the trichromatic theory and its physiological basis. But these are historical questions which are outside the scope of this book. I attack the notion of 'a sensation' in connection with visual perception in Chapter 9, but I do not try to disentangle the thought that *colour* might be *a* sensation.

The interpretation of Goethe is also important because the definitions I give under (2) are derived from a modern interpretation of the *Theory of Colours*.[12] Until Chapter 7,

[11] James Clerk Maxwell, 'On Colour Vision', in David L. MacAdam (ed.), *Sources of Color Science* (Cambridge, Mass: M.I.T. Press, 1970), p. 75.
[12] Due to Wilson and Brocklebank: see Chapter 7.

however, I shall not, strictly, be discussing colours, but rather the being coloured of coloured things, these things further restricted to physical bodies. This distinction, for reasons of exposition, is not always marked in the text. In Chapter 7 there is a purification. My concern is principally with object colours (chemical colours in Goethe's sense), not with physical or physiological colours, or colour effects. The type of answer which I give to the puzzle questions using the definitions also invites the brilliant new view of the senses given by J.J. Gibson in *The Senses Considered As Perceptual Systems*,[13] and his Aristotelian concept of experience as the extraction of significant content from the environment, not as the presentation of matrices of indefinable sense-data. Gibson argues that perception does not depend on sensations, a point which I endorse in Chapter 9. I would also say that our received conception of a sensation is defective.

If Gibson is right, however, then it should come as no surprise that the meaning of colour terms cannot be fixed by reference to sensations. There is in the text an argument concerning the meaning of colour terms which I have not tried to develop. What I would want to argue if space permitted is that the view of colour which is required for the solution to Wittgenstein's puzzles itself requires the Leibniz–Putnam–Kripke theory of reference. Alternatively, perhaps, what I have shown is that 'white' and the other colour terms are, or are very like, natural kind terms in Putnam's sense. I must confess that as I wrote Chapters 2–6 it never occurred to me that they were *not* natural kind terms in this sense. What does emerge clearly is that white is not or is not only a 'phenomenal property': or rather it would mean that, if this term of art had ever received a satisfactory definition in the literature. It is a significant feature of the Leibniz–Putnam–Kripke theory that the real essence which is invoked in the theory of reference is separated from the a priori. If the method I have adopted is the right one, then none of the puzzle propositions are known a priori to be true, but all of them are necessary. Insofar as

[13] (Boston: Houghton Mifflin, 1966.)

the idea of epistemic necessity depends on the a priori, the puzzle propositions are therefore not epistemically necessary. This suggests a weakness in the very idea of epistemic necessity, since the puzzle propositions might be thought to be epistemically necessary if any propositions were. And insofar as epistemic necessity is understood merely in Dummett's sense as knowability a priori,[14] it will also suggest a weakness in the very idea of the a priori.

1.5 Final Remarks

It is a worrying feature of the text of *Remarks on Colour* that Wittgenstein nowhere gives specific answers to any of the puzzle questions. Neither does this appear to have disturbed him or to have engaged his doubts about his own methods. Perhaps he really did believe that we should somehow work free even of questions as specific as the puzzle questions, rather than find answers to them. When he wrote at *Philosophical Investigations* 109 that 'We must do away with all *explanation*, and description alone must take its place', he meant, or should have meant, by 'explanation' one particular kind of constricting half-truth, which in our culture has come to be accepted as virtually equivalent to explanation. So he might in the end have been sympathetic to the attempt made here to find *answers* which are simultaneously attempts to dissolve the puzzle questions, i.e. to do *complete* justice to the phenomena which prompt them. I am not sure whether I have advanced a theory in the pejorative Wittgensteinian sense in this monograph. If I have, I hope that nothing it contains will make Wittgenstein's questions appear less interesting or less important than they are.

[14] Michael Dummett, *Frege: Philosophy of Language* (London: Duckworth, 1973), pp. 115 ff.

2

White

> Bethink thee of the albatross: whence come those clouds of spiritual wonderment and pale dread, in which the white phantom sails in all imaginations ... Nor, in some things, does the common hereditary experience of all mankind fail to bear witness to the supernaturalism of this hue ... Is it that by its indefiniteness it shadows forth the heartless void and immensities of the universe, and thus stabs us from behind with the thought of annihilation, when beholding the depths of the milky way? Or is it, that as in essence whiteness is not so much a colour as the visible absence of colour, and at the same time the concrete of all colours?... And of all these things the Albino whale was the symbol. Wonder ye then at the fiery hunt?
>
> HERMAN MELVILLE, *Moby-Dick*, 'The Whiteness of the Whale'

> The truly best therapy is a sensible theory of the world.
>
> HILARY PUTNAM

2.1 The Main Problem

Why is it that something can be transparent green but not transparent white?[1] The answer to this question and to the other puzzle questions which Wittgenstein asks in *Remarks on Colour* does not, he claims, belong to the physics, physiology or psychology of colour, but to something he calls the 'logic of colour concepts'. The relevant sciences cannot answer the question as he means it, Wittgenstein says.[2]

[1] *Remarks on Colour*, I 19.
[2] Ibid., I 39.

Exactly how he means it, however, is something which is only fixed when we see what he countenances by way of an answer. Unfortunately the answer he gives is unclear, or he doesn't give an answer at all. I cannot find one in the text of *Remarks on Colour*. I hope that the answer which I give to Wittgenstein's puzzle question is a clear one, even if it is misguided and I have not understood the point of the question as Wittgenstein means it. But if my answer is the right one, then science does bear upon the logic of colour concepts, and the distinction between logic and science which Wittgenstein sets up is a false one. At best it will mark a contrast between the demands of logic, and the claims of a particular scientific theory and mode of scientific theorizing.

2.2 A Confusion

Before starting on the question itself it is necessary to remove a small but important confusion. In his review of *Remarks on Colour*,[3] Nelson Goodman claims that Wittgenstein's question is 'mistaken'. He points out that 'the glass in a white light bulb sometimes is as transparent as that in a red one.' This is true; but it is also not to the point. 'As transparent as' does not mean 'transparent', any more than 'as full as' means 'full'. Two jugs which are not full can be as full as one another, e.g. half-full.

Goodman has confused his true proposition, 'A white glass can be as transparent as a red one' with a different and false proposition, 'A white glass can be transparent.' According to the *OED*, 'transparent' means 'having the property of transmitting light so as to render bodies lying beyond it completely visible, so that it can be seen through.' Goodman's white light bulb is not transparent, it is merely *translucent*. Translucency is partial or semi-transparency, and 'translucent' means 'allowing the passage of light yet diffusing it so as not to render bodies lying beyond it clearly visible.' Goodman's white (pearl?) bulb can be seen not to be transparent by comparison with a

[3] Nelson Goodman, in the *Journal of Philosophy* lxxcv 9 (1978), p. 504.

completely transparent or colourless bulb, in which the filament is clearly visible. A white bulb can have the same degree of translucency as a red one, but for it to be as transparent as the red what lies behind it must be as clearly visible as it is through the red. (The bodies lying beyond or behind a transparent glass must be visible as normal, so that the fact that bodies flush against a white glass are somewhat visible is not enough to make the glass transparent. What is actually seen in such a case is a shadow on the glass. The *OED* definition requires that objects lying *beyond* the transparent medium, not merely behind it, must be completely visible.) Perhaps we should say that no coloured medium is transparent in the strictest sense that it transmits all the light incident upon it, and renders any body beyond it as visible as it would be without the interposition of the coloured medium. But even if no body is fully transparent in this sense, why is white glass *less* transparent than red? The fact that *a* white glass can be as transparent as *a* red glass, that is, as transparent as *some* red glass, does not show either that white glass is as transparent as red or that white glass is or can be transparent. Some white glass, the most nearly transparent, has the same degree of transparency as some red glass – relatively untransparent red glass. So Wittgenstein's question is not mistaken. Goodman's answer (there can be a transparent white glass because there is) is itself mistaken. Moreover, even if it were not, it would leave over a variant of Wittgenstein's question: why is white glass generally less transparent than glass of any other colour? Why are there no white-tinted spectacles?

2.3 White is a Surface Colour

There is in fact a crucial difference between white and the colours which do allow transparency. White is typically a surface colour, not a film or volume colour.[4] David Katz

[4] David Katz gives an account of these distinctions in *The World of Colour* (London: Kegan Paul, 1935), p. 7, and see also C.J. Bartleson, Robert W. Burnham, Randall M. Hanes, *Color: a Guide to Basic Facts and Concepts* (New York: Wiley, 1963), pp. 50–51. Colours are said to appear in surface, volume, film or aperture, illumination and illuminant modes.

reports observations by Gelb concerning the loss of perception of surface colour by a patient as a result of an occipital lesion. 'The patient was unable to localize the colours of objects' at precise distances, 'colours appeared to the patient to have a spongy texture', 'they failed to lie flat on the surface of objects', and 'the patient had to reach *into* the colour in order to touch the surface of the coloured object. He had to plunge farthest in when the paper was black and least when the paper was white.'[5] This patient *never used the terms 'black' and 'white' to report his non-surface colours; he always used 'bright' and 'dark' instead.* Katz says of surface colours,[6] which David Wiggins has usefully labelled *barrier* colours, that paper coloured these colours

> . . . has a surface in which the colour lies. The plane in which the spectral colour is extended in space before the observer does not in the same sense possess a surface. One feels that one can penetrate more or less deeply into the spectral colour, whereas when one looks at the colour of a paper the surface presents a barrier through which the eye cannot pass. It is as though the colour of the paper offered resistance to the eye.[7]

Why should this be, and what exactly is the 'resistance' which white, more than any other colour, offers to the eye? Is it right to say that it is the *colour* rather than the surface which offers the resistance? And why is white a colour which characteristically appears in the surface mode?[8]

[5] Katz, p. 14.

[6] It is surely a grammatical mistake to speak, as Katz does, of surface, film, and volume *colours*. Blue, for example, can be all three. It would be better to say, as Katz also does (I believe following Husserl), that these are modes of the appearance of colours rather than modes of colours. Cf. *Remarks on Colour*, III 202: 'It is odd to say that white is solid, because of course red and yellow can be the colours of surfaces too, and as such, we do not categorically distinguish them from white.'

[7] Katz, p. 8.

[8] It is also said to appear in the illuminant mode, e.g. as the colour of naval signals, but never in the illumination mode. For the claim that it does not appear even in the illuminant mode, see Irvin Rock, *An Introduction to Perception* (London: Macmillan, 1975), p. 503. A white spherical lighting fixture is in a sense a white light (it is a lamp), and it looks white or is white to look at. But this does not contradict Rock's view, for what is white is the illuminated white plastic of which the lamp is made. Rock's view is that an incandescent bulb 'may appear to

Wittgenstein says that 'Opaqueness is not a *property* of the white colour. Any more than transparency is a property of the green.'[9] It is not clear what Wittgenstein intends to contrast here with the accented 'property', and I shall take him to mean that the *colour* green, if we can put the matter this way, is not what is transparent, although it must be conceded that the accent on 'property' rather than 'colour' makes this an improbable reading.[10] So what is it that is transparent or opaque? Wittgenstein's remark prompts the thought that transparency and opacity are properties of objects, substances, media, etc. – milk, glass, paper, perspex, varnish, water, cotton, magnesium oxide, and so forth – and not properties of the colour itself. This seems to be correct. I shall have something to say about why it is later.

As with the other impossible colours discussed in *Remarks on Colour* (reddish green, pure brown, glowing grey, etc.), Wittgenstein's interest is fixed not on the mere nonexistence of transparent white, in the sense in which mauve did not exist before Perkin synthesized and christened a 'mauve' coal tar dye in 1856, but on its impossibility or inconceivability.

> Why is it that something can be transparent green but not transparent white?[11]
>
> *Why* can't we imagine transparent-white glass, – even if there isn't any in actuality? Where does the analogy with transparent coloured glass go wrong?[12]

Notice the 'something' in the first of these remarks, 'glass' in the second. Wittgenstein also says[13] that we cannot describe (paint) something simultaneously white and clear,

---

be bright or dim, but in neither case would we refer to it as having a colour such as white or grey. Rather it appears to be emitting light, to be shining or luminous. (The term *white light* is, therefore, unfortunate; rather we should speak of *neutral light.*)'

[9] *Remarks on Colour*, I 45.

[10] But cf. *Remarks on Colour*, III 242; 'milk is not opaque because it is white, – as if white were something opaque.'

[11] Ibid., I 19.

[12] Ibid., I 31.

[13] Ibid., I 23.

and we cannot describe how such a thing would look, or should look. This means that 'we don't know what description, portrayal, these words demand of us.' White water and clear milk are 'inconceivable'. (Why would clear cream not be thick enough to *be* cream – creamy? 'Cream' is also a colour term. Is it phenomenology, logic, synaesthesia (psychology) or nonsense that cream is a *thick* colour?)

2.4 Conditions for a Solution

It has been suggested that Wittgenstein wants to *replace* the question why white water is inconceivable etc. with the question why we don't know what the words 'transparent white' demand of us.[14] This seems wrong, even apart from the fact that nowhere in *Remarks on Colour* does Wittgenstein actually *say* that this is his purpose. He doesn't even *ask* the second question, why we don't know what the words demand of us. All he says is that we don't. In fact the second question, the one Wittgenstein doesn't ask, is actually the general form of his *solution* to the puzzle. The point is that there is nothing the words could demand of us – 'We don't know what these words demand of us' is a Wittgensteinian way of saying that they make no coherent demand. Their failure is a *logical* failure in the sense of *Remarks on Colour* I 27: 'When dealing with logic, "One cannot imagine that" means: one doesn't know what one should imagine here.'

It is as though Wittgenstein's method, insofar as he has one, is to try in various ways to imagine the thing of which it is true that 'one doesn't know what one should imagine here'. The knowledge of the inconceivability emerges from these doomed efforts. Thus 'A smooth white surface can reflect things: But what, then, if we made a mistake and that which appeared to be reflected in such a surface were really behind it and seen through it? Would the surface then be white and transparent?'[15] Wittgenstein's use of this type of oblique reductio ad absurdum makes it difficult to

[14] H.O. Mounce, *Philosophical Quarterly* **119** (1980), pp. 160–161.
[15] *Remarks on Colour*, I 43.

piece together a specific solution to the whiteness puzzle, if there is one, in the *Remarks on Colour*. One searches in vain for a straightforward statement of the central features of the concepts of whiteness and transparency whose hidden logical relationship is responsible for the problem. The nearest Wittgenstein comes is the 'rule of appearance' he gives at I 20 and III 173. 'Something white behind a coloured transparent medium appears in the colour of the medium, something black appears black.' Wittgenstein says that this is not a proposition of physics, but rather a rule of the spatial interpretation of our experience, and that it could also be called 'a rule for painters'. But although he derives an absurd consequence from the rule (black on a white background would have to appear through a transparent white medium as though through a colourless one),[16] we do not get a formal contradiction. It would be nice to have one if we are to follow Wittgenstein in speaking of the impossibility or inconceivability of transparent white (it would explain *why* 'one doesn't know what one should imagine' here) for without it we won't get to the bottom of the problem. '*Why* can't we imagine a transparent-white glass?' The spade is turning – do the italics signify desperation?

In order to solve Wittgenstein's puzzle we need to be able to say something about what whiteness and transparency *are*, or (what I regard as the same thing) give the concepts of these properties, or state what it is to be white and transparent, i.e. give a theory of whiteness and transparency. We must be able to do more than identify them by gesturing in their direction and hoping that a word emitted will follow the gesture through to the intended target. We must do what the empiricists have told us we cannot do, namely something which has the function of 'unpacking' the concepts of whiteness and transparency, and resolving them into simpler concepts. If this can be done, and one need not accept the empiricist account of what must be

[16] But then, why shouldn't it?

done in order to do it, then colours (at least one of them) will have been shown not to be logically simple.[17]

An acceptable conception of whiteness will have to connect the concept simultaneously with the phenomenal property (the *colour*, as I would prefer to say) and the physical property which is supposed in psychophysics merely to stand in a correlation to it. It will have to do more than supply a deictically based correlation between the physical magnitude and the so-called phenomenal property, conceived, somehow, as a property or whatever of a psychological or physiological state, or else the problem will reappear in an even darker and more unregenerate form. Why should just this correlation hold?[18] Why should it be phenomenal *white*, and not some other colour, which correlates with maximum impurity in the Fourier analysis sense and a particular pattern of spectral selectivity? Could not the correlation have been with phenomenal *green* instead? Is it merely contingent physiological functions which prevent this?[19]

We need a conception of whiteness which will connect the colour with the unique or asymmetrical property (asymmetrical in that no other colour has it) of necessary opacity which is our explanandum. Opacity is, like transparency, a physical property of objects, substances, surfaces, etc. The conception must not allow the physical property and the phenomenal property to part company, or we will lose the prospect of finding here a necessity, logical or otherwise, if there is one, which is the main element in Wittgenstein's puzzle. I now offer such a conception. It makes clear, I hope, just what the genesis of the logical element in the proposition that nothing can be white and transparent is. I

[17] See Chapter 8.

[18] There is also the more general question, in Chapter 8, why there should be a contingent relation here at all.

[19] It is worth pointing out that here we have a version of the spectrum inversion problem, and that this problem, which derives from the seventeenth century, is powered not so much by epistemological arguments as by a world conception in which the physical world and the phenomenal world are related not as reality and appearance but as cause and effect.

do not, however, regard it as a complete conception. Very much more needs to be said than I say here about whiteness in contrast effects, in adaptation phenomena, and generally in connection with the physiology and psychology of colour perception. I believe that it could be made complete, but this is not necessary for the limited purpose of solving Wittgenstein's puzzle. Here all that is needed is the special case which is the physical part of the theory of lightness. The conception will, I hope, make good some of the large claims already made as to the correct form which the solutions to the puzzle questions should take.

2.5 What Is Whiteness? The Problem Solved

A white surface is a surface which scatters back or reflects nearly all of the light incident upon it. It is a surface with a reflectance of approximately eighty per cent or more across the spectrum, or for white (neutral, colourless) light. If the percentage of reflected light were much less than this, for example if more light was absorbed and transmitted than was reflected, the result would be something other than white. The ratio of the reflectance, absorbance and transmittance of a substance will vary for different coloured lights across the spectrum, but in all cases the sum of the three is naturally the same.[20]

So what would a transparent white surface be? It would be a surface which (i) transmits almost all the incident light – it is transparent – enabling us to see what lies beyond it, and (ii) scatters back almost all the incident light – it is white – and transmits almost none. A transparent white object would transmit almost all the incident light and

[20] Keith Campbell, in 'Colours', in Robert Brown and C.D. Rollins (eds.), *Contemporary Philosophy in Australia* (London: Allen & Unwin, 1969), p. 137, makes the alleged fact that not all objects of a particular colour under a given illumination have 'a distinctive light-modifying feature in common' an objection to the idea that 'colours are intrinsic physical qualities of surfaces'. High diffuse reflectance is common to all white surfaces. It is not relevant that there may be many *causes* of high reflectance. There may be no property common to all square things which makes them square or makes them to be square (apart from being square) but this has no tendency to show that being square is not an 'intrinsic physical quality', however we construe this, of a surface.

eflect almost none of the incident light (it is transparent) nd reflect almost all the incident light and transmit almost one. This is a straight double contradiction. Elizabeth Hackett has suggested that these facts should be repre-ented in the truth-table below, where *W* is the proposition hat something is white, *Tr* the proposition that it is trans-parent, *r* that it has a high diffuse reflectance for light of all olours, and *m* that it has a high transmission for light of all olours.

		W	*T*	*W & Tr*
r	*m*	*r & –m*	*–r & m*	*(r & –m) & (–r & m)*
T	T	F	F	F
T	F	T	F	F
F	T	F	T	F
F	F	F	F	F

Note that *r* and *m* are not strictly contradictory. I assume hat since it is so obviously a logical impossibility that omething cannot simultaneously transmit and reflect all r most of the incident light, 'transmits' and 'reflects' can hemselves be further analysed until we reach a strictly ogical contradiction.

It used to be said that every kind of statement must have s own kind of logic – so then every kind of statement must ave its own kind of contradiction. But in the proposed onception '*X* is white and *X* is transparent' turns out to be n ordinary contradiction of the form p & −p. So there is ere no peculiar or distinctive logic of colour, but only logic pplied to ordinary statements about the distinctive facts nd phenomena of colour.

A white surface, then, or a white object, medium or ubstance, will always reflect most of the incident light, and herefore it will not darken the light to any significant egree, by absorbing some of it, as surfaces of other olours will. A red object, for example, refuses to reflect ny green light, and so turns black in green illumination. A white object darkens no light in the sense that for any llumination colour the incident light is approximately the ame in quantity and quality as the reflected light. So white is always the lightest colour in the Tricolour,[21] not because the colour concept 'white' inexplicably happens to be: the concept of something with the property of being the lightest colour (why should it?), but because the concept of a white object or surface is the concept of something which does not significantly darken the light in the above sense. (Not: 'is the concept of: something which' but 'is the concept: of something which'.) Actually the better question is not why white is the lightest colour, but rather why all other colours are darker than white. In Goethe's conception of a colour the answer is that black and white are limiting cases of the darkening process which for him is what 'physical' and 'chemical' (object) colours are.[22] White is minimum darkening, black is maximum darkening.[23] White is 'the representative of light', in his classic phrase.[24] We can be

[21] A fact which Wittgenstein uses to introduce the 'sort of mathematics of colour' at *Remarks on Colour*, III 2.

[22] Cf. *Remarks on Colour*, I 52: 'White as a colour of substances (in the sense in which we say snow is white) is lighter than any other substance-colour; black darker. *Here* colour is a darkening, and if all such is removed from the substance, white remains, and for this reason we call it "colourless"'. Cf. W.D. Wright, 'Towards a Philosophy of Colour', in *The Rays are not Coloured* (London: Hilger, 1967), p. 26, on the use of the concept in which 'when a painter talks about colour he means pigment.' 'A colour' can mean a paint, and painting on white paper is darkening it. For a sympathetic modern exposition of Goethe's theory, M.H. Wilson and R.W. Brocklebank, 'Goethe's Colour Experiments', in the *Year Book of the Physical Society* (London: The Physical Society, 1958).

[23] A limiting case: white might be regarded as a colour in the same sense in which zero can be regarded as a number (the answer to, 'How many?' can be 'None'), but also as the absence of a number or number.

[24] *Theory of Colours*, I, 1, 18, p. 7. Goethe was the first to reassert or reinvent the pre-Newtonian claim that colour is an edge-phenomenon, which was conclusively established by experiments with absolutely stabilized images in the 1950s. See 'A Necessary Edge', in 'Confessions of a Color Enthusiast', *Journal of Color and Appearance*, Vol. 1, No. 3, Nov.–Jan. (1972): 'It required little thought to realize that an edge was necessary to bring about colour.' An edge could be regarded as a differential darkening. For an introduction to the edge-phenomena, see Lloyd Kaufman, *Perception: The World Transformed* (New York: Oxford University Press, 1979), Chapter 5, 'Of Patches, Bars and Edges'. Newton dismissed the claim in Isaac Newton, 'Letter to Oldenburg', in *The Correspondence of Isaac Newton*, ed. H.W. Turnbull, Vol. 1 (Cambridge: Cambridge University Press, 1959), p. 92: 'I could scarce think, that the various *Thickness* of the glass, or the termination with shadow or darkness, could have any influence on light to produce such an effect' (a spectrum), and he repeated his dispersion experiment with 'holes in the window of divers bignesses . . .' For a discussion of Goethe's

assisted in this phenomenal conception by watching a white surface turn under increasing intensity of the light source first to glare and then to dazzle, which lie higher on the brightness scale.[25] There is such a thing as a glaring white and a dazzling white, but no such thing as a glaring black or a dazzling black. Why? Other similar facts may bring home the *phenomenal* rightness of the conception. Why, for example, does a painter use *white* for highlights? We might even say that whiteness is a *low dazzle* of reflected light.[26]

---

reply, see Neil M. Ribe, 'Goethe's Critique of Newton: A Reconsideration', in *Studies in the History and Philosophy of Science*, Vol. 16 no. 4 (1985), pp. 329–330. The edge theory also contradicts the view that white can be the colour of light or the colour of a light, and restricts it to objects and surfaces. The definition of whiteness I give is standard in its physical interpretation, though since 1967 the perfect diffuse reflector has replaced the smoked layer of magnesium oxide as the standard, and colorimeters are calibrated by sub-standards that are themselves related to the perfect diffuser. An equivalent phenomenalist definition due to Richard Mort makes ideal white surfaces those which darken no light, and therefore do not darken it selectively, i.e. darken light of some colours but not of others. For the standard definition, see e.g. R.P. Blakey and G. Landon on the perfect diffuse reflector, 'An absolute standard of whiteness', and Appendix C, 'The perfect diffuse reflector as a C.I.E. standard', in *Measuring Colour* (London: British Titan Products, 1970), p. 11. The virtue of Mort's definition is that it allows the method used to answer Wittgenstein's puzzle questions about whiteness to be extended to the puzzle questions about the other colours. It also allows us to define 'white' light in a non-conventional way. We cannot say that 'white' light is light none of which is absorbed or all of which is reflected by an ideal white surface, for this is true of light of all colours. But we can say that a white light is one which does not darken an ideal white surface. In practice the standard C.I.E. Illuminants and the newer D_{65} involve conventions in principle for reasons having to do with the arbitrariness of choice of viewing conditions.

[25] Here there is no clash between science and commonsense, or between science and perception. A white surface looks exactly as we would expect a surface diffusely reflecting or unselectively not darkening the incident light to look, insofar as there are any prior expectations here.

[26] I find some confirmation of this view in the *leukos* of Attic Greek, a Stage IIIb language in Berlin and Kay's classification (*Basic Color Terms* (Berkeley: University of California, 1969)), i.e. one which covers colour space with only four terms whose foci or areas of primary application are the same English 'white' ('leukos'), 'black' ('glaukos'), 'red' ('erythros') and 'green' ('khloros'). 'To Leukon' translates as 'white', but also 'light'. The spread of the colour term over colour space and also the connections with other concepts differ. 'Leukon' apparently describes water, sun, metallic surfaces, anything bright, reflecting, clear. It is much less obviously the name of a simple idea than the English 'white'. For some 'obvious inaccuracies' in Berlin and Kay's account of Attic Greek, see Eleanor Irwin, *Colour Terms in Greek Poetry* (Toronto: Hakkert, 1974), p. 24.

2.6 Adaptation

What is dazzling, however, depends on the adaptive state of the eye and therefore on relative rather than absolute brightness. We must distinguish reflectance, the *proportion* of incident to reflected light, from luminance, the absolute amount of light entering the eye. Thus in a famous example a white sheet of paper in shade looks the colour it is (white, not grey), and a grey sheet of paper in direct sunlight which is reflecting far more light looks grey. Whiteness is connected not with the quantity of light entering the eye but with the ratio of reflected and incident light, that is, with what the surface typically *does* to the light. The colours of objects that we see are always perceived in relation to the illumination. The illumination, no matter what colour it is, is treated by the eye as standard or neutral, and the other colours are 'judged' in relation to this. The reflectance of white objects, as contrasted with the luminance, is the same through changes in illumination, and therefore we get as a natural result what psychologists call colour constancy. This would be a properly psychological effect only if the important property for colour vision was the spectral composition and absolute quantity of light entering the eye. But the eye is not concerned with the colour and quantity of light. It is concerned rather with how the object *changes* the light.[27]

[27] This principle also applies to brightness. 'Due to the powers of adaptation of the eye to varying conditions of illumination, there is little connection between the apparent brightness of a surface and the absolute intensity or quantity of reflected light. It is the reflected *fraction* of the incident light which is all important' (Frederick W. Clulow, *Colour: Its Principles and Applications* (London: Fountain, 1972), p. 24. As regards lightness, I mean to take sides in the complex and fascinating debate between the classical theory of lightness constancy due to Helmholtz, in which the illumination is taken into account, and the modern ratio theories deriving from Hering. For an account of this discussion, as well as helpful remarks on the confusing distinction between lightness and brightness, Irvin Rock, *An Introduction to Perception*, pp. 515 ff., 'Two Theories of Neutral Color and Neutral Color Constancy'. David Katz writes in *The World of Colour*, p. 1, that 'Whether objects are located in the immediate vicinity or miles away from us, we always perceive between ourselves an empty space in which no objects are to be found ... With the same immediacy with which we perceive the colours of objects comes the apprehension of their illumination; and the illumination is not

A white surface scatters back almost all of *any* incident light, light of any colour. This is not true, for example, of a blue object, which under a blue light will do exactly what the white surface does. Provided we are aware of the illumination, we can tell the difference between the two objects, even if there is no physical difference between them. This is because their relation to the illumination over the whole scene is different. If on the other hand we are prevented from adapting to the illumination, we cannot tell the difference between the colour derived from the illumination and the colour which is a function of the chemistry of the object. We cannot tell the difference between a blue room filled with blue objects and illuminated by white light, and a white room filled with white objects illuminated by a blue light *if we are looking in through a small window*. Under this restriction we cannot tell in what characteristic way the objects in the room *change the light*. It should be pointed out, against most philosophers who have discussed this sort of example, that a white surface, say a wall, in blue light does *not* look blue – provided we are able to adapt to the colour of the illumination. It looks *white*, and completely different from a blue surface illuminated with white light.

This mistake is made by Ayer and many other writers.[28] One of the empirical premises on which the argument from

---

limited to the objects alone, for the empty space between objects is also seen as illuminated.' In *The Perception of Colour* (London: Wiley, 1974, p. 7), Ralph M. Evans comments that this 'immediately eliminates the possibility of a one-to-one relation between the local physical stimuli and the perceptions they produce'. The fact of the perception of the illumination provides a link between the perception of colour, which is only apparently spatial (see Ch. 7 of the present work), and the perception of space.

[28] A.J. Ayer, *The Central Questions of Philosophy* (London: Weidenfeld, 1973), p. 74. The mistake appeared in identical form in R.E. Tully, 'Reduction and Secondary Qualities', *Mind*, lxxxxv **339** (1976), p. 352, and in a similar form in Timothy L.S. Sprigge, *Facts, Words and Beliefs* (London: Routledge, 1970), pp. 4–5. For some reason philosophers seem to want to choose a *white* wall seen through *blue* spectacles, which is easily the worst example for the point they wish to make, since the blue actually enhances the white by making up the blue deficiency of many white objects. (Optical bleaches are blue-white fluorescent, and laundries use blue crystals to make their whites whiter than white – which is a real effect (note the dip of the curve for typing bond in Fig. I (p. 32) in the blue).

illusion (to sense-data) is based says that a white wall 'looks blue when it is seen through blue spectacles'. This premise is simply false. (Does Ayer believe it looks black through black spectacles?) The wall continues to look utterly white through blue spectacles, not blue, although as if in a dimmer and slightly misty illumination, as it does in blue illumination. If on the other hand we place the spectacles up against the wall, we do get a blue datum, but one which is obviously the colour of the spectacles. When we are wearing the spectacles, we adapt to what is effectively the colour of the illumination. When the blue is not the colour of the light but of the spectacles held up against the wall, we are aware of how they change the now white illumination to which we have adapted. (Also the light is darkened twice, once on the way to the wall through the spectacles and once coming back, so the colour is twice as strong.) Wittgenstein's crucial 'rule of appearance' ('White seen through a coloured glass appears with the colour of the glass') is simply false.[29] It applies only in non-adaptive situations. Accordingly, it cannot explain the opacity of white things or anything else, and Wittgenstein's deduction fails.

2.7 More Puzzle Propositions

Several Wittgensteinian explananda fall into place under the proposed conception.

(i) We learn why white is a surface colour, and reverts to achromatic brightness in other modes. In order for anything to be white it must scatter back the required percentage of the incident light. If it can do this, it is probably going to be sufficiently *solid* to count as a surface and present a literal barrier to the light.

(ii) It becomes evident why a white lying behind a coloured medium in a non-adaptive situation, or when the effective

[29] Wittgenstein does say, as we saw, that the rule of spatial appearance is not a proposition of physics but a painter's rule. But he also says (III 180) 'We are not concerned with the facts of physics here, *except insofar as they determine the laws governing how things appear*' (my italics). There is the question *why* the painter's rule is true – when it is.

illumination is so strong that we cannot adapt to it, should appear with the colour of the medium. The white surface acts as an achromatic light shining through the coloured medium, which functions as a simple filter.

(iii) *Remarks on Colour* II 16 is confirmed: 'Isn't white that which does away with darkness?'

(iv) 'We don't speak of a "whitish cast on things" at all!'[30] This is a very important and interesting observation. We may be prepared to speak of white lights, a white light, etc., as illuminants or sources of illumination, but 'a whitish cast' in this remark means something different. We might notice that the light at a particular time of day was bluish, but not that it was whitish in the sense that it gave a whitish tinge to things. The distinction to be drawn here is between a light in the sense of a lamp and light as the illumination or way of being lit. The former can be phenomenally white, the latter not. We have defined a white surface as a certain alteration of the light by a surface. So 'if everything looked whitish in a particular light, we wouldn't then conclude that the light source must look white',[31] and still less that the actual light from the source must look white. It could only be a change in the surfaces of things which made them whitish. The illumination never looks the same colour as snow – and this is a good thing, because otherwise we would not be able to see through it, like living in a permanent blizzard. Light in this sense is not white *coloured*. It is only white in a specialized sense. Newton says, in the *Opticks*, Prop. VI, Problem I, 'I placed a lens by which the image of a hole might be distinctly cast upon *a white sheet of paper* ...' (my italics). A light which gives red on a white sheet of paper must be a red light, a light which gives blue on a white sheet of paper must be a blue light, so a light which gives *white* on a white sheet of paper must be a white light.[32] But why did Newton choose a white sheet of paper? The white given by the 'white' illumination is the

[30] *Remarks on Colour*, II 14.
[31] Ibid., II 15.
[32] 'In the course of a scientific investigation we say all kinds of things: we make

colour *of the paper* already. It 'appears' because the 'white' light is strictly colourless (it makes red things red), and the screen is already white. When we say such things as 'The light is poor today' we mean light as *lighting*, or perhaps the way of being lit. Light in this sense cannot appear white, like a white mist, and nor does it make anything white except what is already white. Rather, it allows the colours things already have to be seen. Nor can light in this sense be passed through a prism, though a beam of light can. A beam of light is made of light, perhaps, but it is *a* light rather than lighting.

(v) '"Transparent" could be compared with "reflecting"', Wittgenstein says cryptically at III 148, and he goes on to point out that both transparency and reflection exist only in the dimension of depth of a visual image. The fact that white surfaces present a barrier to the light means that they block the dimension of depth required for transparency of the image.

(vi) 'Blending in white removes the *colouredness* from the colour; but blending in yellow does not. – Is this the basis of the proposition that there can be no clear transparent white?'[33] It is hard to see why it should be. And what does it mean to say that blending white into a colour C removes the colouredness from C? (There is presumably an exception if C is white.) Blending progressively more red into yellow finally destroys the yellow, just as blending in white does. The difference is that the final result in the case of white is the absence of colour. Blending white into C removes C and leaves something colourless because here white is not counted as a colour. We reserve a special

---

many utterances whose role in the investigation we do not understand. For it isn't as though everything we say has a conscious purpose; our tongues just keep going. Our thoughts run in established routines, we pass automatically from one thought to another according to the techniques we have learned. And now comes the time for us to survey what we have said. We have made a whole lot of movements that do not further our purpose, or that even impede it, and now we have to clarify our thought processes philosophically' (L. Wittgenstein, *Culture and Value* (Oxford: Blackwell, 1980), p. 64e).

[33] *Remarks on Colour*, II 2.

concept for the effect of white in colour mixing. White has the effect of making other colours paler, weaker, or less 'saturated' with colour. There is no corresponding concept for the effect of any other colour. (Black has a similar asymmetrical property. It makes every other colour darker.) Everything turns, then, on the question why 'white sometimes appears on an equal footing with the other pure colours (as in flags) and then again sometimes it doesn't'.[34] The basis of the fact that white is sometimes counted as a non-colour and that it destroys colour in the sense of hue are illustrated in Fig. II.

Fig. I shows how much of the illumination of a given colour the specified substance will reflect. If the illumination is blue, for example, the vermilion pigment in Fig. I will reflect less than ten per cent, whereas if the illumination is red the vermilion pigment reflects nearly as much of it as the white typing bond paper does. This is why red and white things are indistinguishable in a red light to which we do not adapt. Fig. II shows an ideal white substance, whiter than magnesium oxide or freshly fallen snow, which reflect only about ninety-five per cent of the light. The ideal white substance absorbs none of the light; it is the perfect diffuse reflector. The curves for the other substances range between light and dark. Let us imagine the curves of Fig. I gradually being transformed into the ideal 'curve' at 100 per cent reflectance of Fig. II. The colour of the substances before this transformation depended on their characteristic absorption or darkening pattern across the spectrum. When the substances are whitened they lose this characteristic selectivity.[35] Ideal whiteness is the theoretical limit of the process which gives substances their colour. Chromatically coloured objects change both the quantity and the colour of the light. White objects do neither. In this respect they are comparable to transparent objects. *Not* changing the light can count as a very special way of changing the light, just as zero can count as a special kind of number. (One could also

[34] Ibid., III 211.

[35] Fig. I suggests an easy explanation of the fact that 'white gradually eliminates *all* contrasts, while red doesn't' (*Remarks on Colour*, III 212).

Fig. I

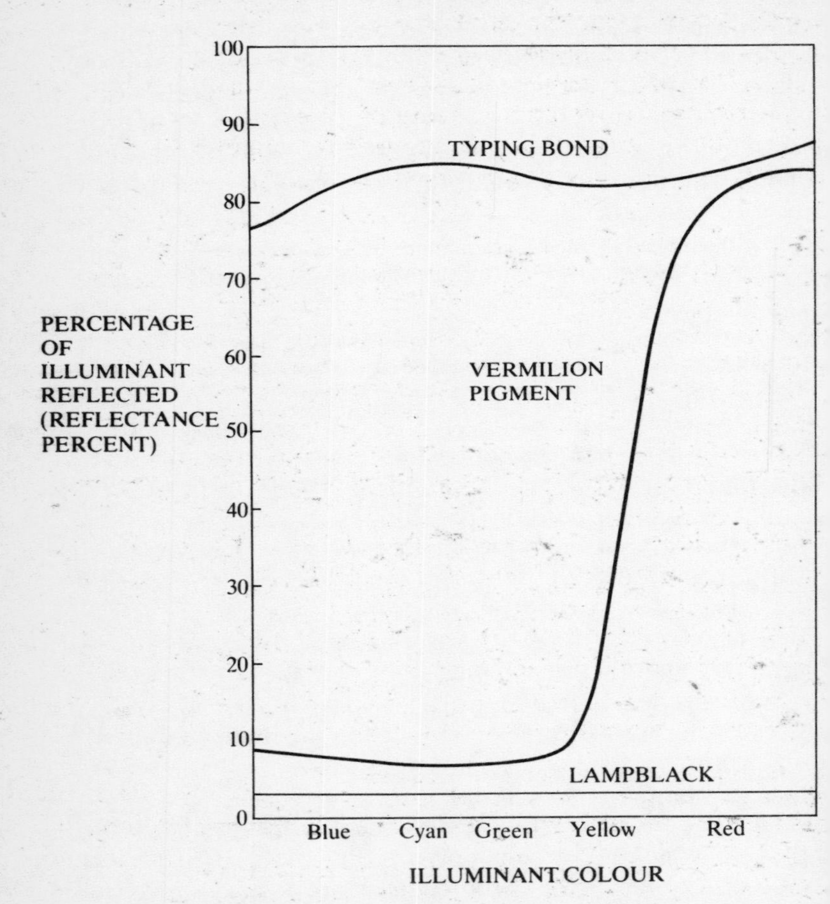

SPECTROPHOTOMETRIC CURVES FOR THREE SUBSTANCES

From Frederick W. Clulow, *Colour: Its Principles and Applications* (London: Fountain, 1972), Plates 18 and 19

Fig. II

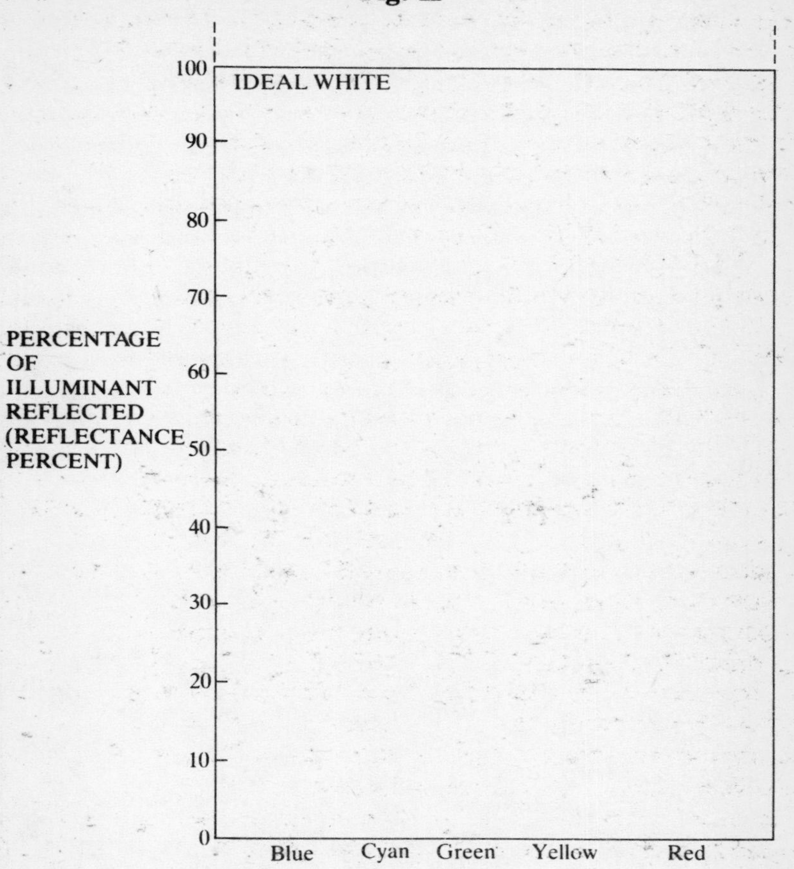

SPECTROPHOTOMETRIC 'CURVE' FOR IDEAL WHITE SUBSTANCE
(THE PERFECT DIFFUSE REFLECTOR)

regard zero as the 'ground' or 'concrete' of all numbers in the sense that putting *four* apples into a box containing zero apples – an empty box – gives *four* apples. Analogously a colour C on white is C.)

The limit of the processes which yield chromatic object colour can also be regarded as the final disappearance or limit of colour, and as an *absence*. How do we choose when to say that white is a colour and when to count it a non-colour? My suspicion is that white is counted a non-colour when we are concerned with *processes* and *changes* in colour – in painting and dyeing for example, where white plays an active role and its special properties have effects. When it is not involved in these processes, white stands as just another 'inactive' object colour. As the colour of paper it already becomes a non-colour, as colour will be placed on it, whereas as the colour of an automobile it is just one colour among others. *Looking* at white will reveal none of these special properties.

2.8 The Right Level of Explanation

All of what has been said so far is quite independent of the theory of what light is. The proposed explanations are consistent with a theory which says that light is 'a Company of little Tennis-balls, which Fairies all day long struck with Rackets against some Men's Fore-heads',[36] wavicles, photons, ectoplasm or nothing at all, provided we end up with the right properties of light, e.g. that it illuminates, is reflected, makes shadows, can be adapted to, etc. A physicalist reduction of colours to light-emissions of different wavelengths, such as David Armstrong has proposed in order to solve the colour incompatibility problem,[37] is

[36] Locke, *An Essay Concerning Human Understanding*, III, iv, 10. The operative point here is not as Locke thinks that the Cause of the Idea of Whiteness could be Globules or anything, and that it doesn't matter what it is in the sense that our Idea of Whiteness would be unaffected no matter what it was, but that it matters very much, and that there are very severe restrictions on what it could be. Locke almost admits this at *Essay*, IV, ii, 2, where he seems to argue a priori that more particles means whiter, and perhaps faster ones mean lighter.

[37] *A Theory of Universals* II (Cambridge: Cambridge University Press, 1978), pp. 126–127. See Chapter 6 of the present work.

simply *irrelevant* to Wittgenstein's puzzle. It selects quite the wrong kind of physical property, and it could not explain the opacity of white things. The important point here is to be found in Hilary Putnam's Pythagorean discussion of why a rigid square *harmonia*[38] won't go into a round hole. The right answer is geometrical (it won't fit), not quantum electrodynamical, an explanation of a *kosmiotes* of stuff – any stuff. It is not relevant, says Putnam, what the peg is made of (atoms, spirits or wood),[39] just as it is not relevant what the string of a monochord is made of; it will give a harmonic octave when halved. The explanation is autonomous in Putnam's sense. The spectral composition of the light reflected from a white surface is not relevant to the question why it cannot be transparent. It is not even relevant to the question whether the surface is white. The right or best logical level of explanation is fixed by what *is* relevant to these questions. It determines and is determined by what kind of thing ('logical grammar') whiteness is. To know what 'light waves are actually being emitted at the surface'[40] is not to know anything which is relevant to the opacity of that surface. As we saw, blue and white things can be reflecting light of exactly the same spectral composition, and yet they are entirely different colours because their relation to the illumination is different. What is relevant to the solution of Wittgenstein's puzzle is that a white surface will reflect *any* light, light of any colour, like a mirror. (Why would it be wrong to say that a mirror image or specular reflection is a light-emission from the surface?) There is then an important sense in which the whiteness we see is not complex. To say that high reflectance for all coloured lights is complex in the sense of being composed of anything is nonsense. Reflecting a high proportion of the light is what the white object does to the light,

[38] A peg (*Odyssey* V 248), according to Guthrie.

[39] Hilary Putnam, 'Philosophy and our Mental Life', in *Mind, Language and Reality*, Vol. 2, (Cambridge: Cambridge University Press, 1975), p. 296.

[40] D.M. Armstrong, 'Colour-Realism' in Brown and Rollins, *Contemporary Philosophy in Australia*, p. 125, thinks that this is what colours must be because 'surely colour is an intrinsic and not a relational property.'

and this doing could not intelligibly be said to be composed of anything.

Consider again Fig. I. It differs from the standard representation of photometric curves in that the x-axis is illuminant colour, not wavelength. This brings out the significant fact that the curves can be obtained by direct observation, independently of the wavelength theory. The vermilion pigment, for example, placed successively in the bands of a spectrum projected onto a screen, will turn very dark in the blue/cyan/green bands and then suddenly very bright in the yellow/red. This is the principle of the spectrophotometer which is used to obtain the curves of Fig. I. It records how much light is reflected by the samples in different illuminations. The older visual spectrophotometers relied on the human eye and human judgements of brightness. They have been replaced by modern photoelectric spectrophotometers, which use a photo cell whose current gives a measure of the quantity of light reflected by the sample. The wavelength values provide a convenient metric for these purposes, but nothing more. The principles of spectrophotometry *are independent of the wavelength theory*. This means that there is nothing in Fig. I to stop us adopting a purely phenomenal interpretation of the whiteness curve, since, as Wittgenstein observes,[41] 'I don't *see* that the colours of bodies reflect light into my eyes', so we can define a white object as *one which does not darken the light* – by absorbing it – and this will be equivalent to defining it by the reflection spectrum. But it is not the same thing for the eye.

2.9 Some Qualifications

Some final caveats about the proposed conception are needed. As it stands it makes metallic surfaces and mirrors white. This may not be such a bad thing (it would not be a bad thing in Homeric Greek phenomenology), but to prevent it all we need is the distinction, already alluded to, between specular and diffuse reflection. In specular or

[41] *Remarks on Colour*, II 20.

directional reflection the image is preserved.[42] There is obvious sense to a concept which will bring out the connections between 'white' and related non-colour concepts such as 'mirror', 'sparkle', 'dazzle', 'brilliance', 'glitter', 'light', etc. There are also good reasons for emphasizing the dissimilarity between white and the related physical phenomena, and the similarity between white and the other colours. As far as I can see there is nothing to prevent us retaining both the wider and the narrower concept, provided we know when we are using which, and why.

Nothing I have said so far shows absolutely that whiteness or the being white coloured of a body is the very same thing as its disposition to diffusely reflect a high proportion of the light, or, in the phenomenal equivalent, not to darken the light. In fact the theoretical definition is not complete, and it will not serve in a physicalist reduction. It does not handle whiteness in contrast effects (e.g. the Gelb effect), though I believe it is the basis of what needs to be said about that from the physical point of view, and clearly it should be stated in terms of relative as well as absolute reflectance values.[43] The theory of the physical nature of whiteness has to be in harmony with the physiological and psychological theory of colour – what needs to be said is something more subtle than that contrast effects are a species of illusion. But the definition is a true part of the complete story. It easily handles the Wittgensteinian aporiai, which must be some kind of evidence, because they are puzzles about the whiteness we actually see. It is visible whiteness which is necessarily untransparent, visible whiteness which is a property of physical objects and surfaces. (In a world of film colours, such as sensations appear

[42] Richard Westphal has suggested that the distinction between specular and diffuse reflection can be phenomenalized by defining a specular reflection as one in which the image is preserved.

[43] How can it be stated in both sets of terms simultaneously? The answer to this exciting question would involve an account of the relation between the metrical physical concepts involved in selective absorption or darkening, reflectance, etc., and the functional relational concepts involved in the concepts employed in the theory of adaptation. I understand 'adaptation' in Helson's sense (Harry Helson, *Adaptation-Level Theory* (New York: Harper and Row, 1964)), in which it involves a heightened performance or sensitizing as well as a desensitizing.

to be, there is no whiteness, only achromatic brilliance.) So the explanation of Wittgenstein's puzzle is a physical one – but not in the low-level reduced sense of physicalism.

And if the concept (theoretical definition, theory, essence) of whiteness did not explain the necessary truths that it generates, I cannot see that anything could. If the *Remarks on Colour* are anything to go by, the successors of 'grammar' in Wittgenstein's *Remarks on Colour* (which gets only one mention in the text, in a question[44]), the 'geometry' and 'mathematics' of colour, the 'logic of colour concepts', are a pretty poor bet. But these fragile metaphors did enable Wittgenstein to frame some of the most suggestive and far-reaching questions that can be asked about colour, both in connection with whiteness and with the other colours.[45]

2.10 Replies to Objections

To complete the solution to Wittgenstein's central problem, I want to discuss three interrelated objections to the type of solution I have been offering. These objections raise difficult questions in the philosophy of language and the philosophy of mind, and I can do no more than suggest where the answers may be found.

(1) The first difficulty I want to discuss is the suggestion that transparent white objects and the other impossibilia of the *Remarks on Colour* would be unimaginable in Wittgenstein's logical sense ('One cannot imagine that' means: one doesn't know what one should imagine here), even if the relevant properties (reflectance, transmittance, light, quantity of light, etc.) were completely different in the sense that they stood in completely different relations to

[44] At *Remarks on Colour*, III 309, 'Here it could be asked what I really want, to what extent I want to deal with grammar.'

[45] My solution to the whiteness-transparency puzzle has been criticized by Colin McGinn in *The Subjective View*, p. 36. Larry Hardin defended my view in 'A Transparent Case For Subjectivism', *Analysis*, **45** No. 2 (1985), pp. 117–119, and Dean Buckner replied to Hardin, defending the subjective view, in 'Transparently False: Reply to Hardin', *Analysis*, No. 8 (1986).

one another. Connected with this difficulty is another, as follows.

(2) 'Strict identity between properties here seems unattainable. We can think of possible worlds in which the phenomenal properties are not associated with the physical properties on which they now rest.'[46] Whiteness might not have been, or might now in some possible world not be correlated with high reflectance across the spectrum, or the disposition of white objects not to change the colour or quantity of the incident light, but e.g. with zero reflectance and high transmittance. If so, then transparent white things would be imaginable in the world – not by us, perhaps, but by its inhabitants, who are used to dealing with the improbabilities and impossibilities inflicted on their world by Earth philosophers. (But what would they understand by 'transmittance'?) The source of the unimaginability has not after all been located. Wittgenstein's question has not been answered, and indeed it could not be answered, as Wittgenstein himself claims, although for different reasons, by the application of physical concepts. I shall make three points about this first pair of difficulties.

(a) First, *could* the relations between the concepts of whiteness, reflectance, transparency, transmittance, surface, reflected light, etc., have been or be in some possible world completely different? This is a very general question, and it is hard to discuss it intelligently without knowing exactly what changes are contemplated. One thing is clear: to the extent that the changes contemplated were very great, there would have to be a commensurate doubt as to whether it really was the concepts of whiteness, reflectance, transparency, etc., that were being talked about. Received colour theory would suffer an upheaval of unimaginable proportions if there were *no* connections between the concepts of whiteness, reflectance, transparency, etc. They would surely be quite different concepts. Making thoroughgoing sense of a world in which no such relations obtained

[46] J.L. Mackie, *Problem From Locke* (Oxford: Oxford University Press, 1976), p. 168. Mackie is discussing the identity of neurophysiological and phenomenal properties, but the logical points at issue are the same.

would leave us unable to make sense of the colour theory we have for the actual world. But of course we can make sense of the colour theory we have for the actual world, and what is more it is precisely this that is the instrument of our conceiving or failing to conceive the allegedly possible worlds. I take this to be a Wittgensteinian point.

(b) So in what sense is it supposed to be the case that we can think of possible worlds in which white things did not have high reflectance, did change the colour and quantity of the illumination, etc? We can *suppose* that there are such possible worlds, but in what sense strong enough to ensure the possibility of the supposedly possible worlds can we think it? We can suppose, imagine, guess and in a weak sense think that some large number is not prime; and for all that it is *necessarily* prime. In order to think of the possible world, in the relevant strong sense, in which whiteness is not correlated with high diffuse reflectance but with some quite other property, we would, it seems to me, have to *think successfully of* or *identify* whiteness in the possible world. How is this to be done? *Mentioning* it is not enough, and thinking of it in the relevant strong sense (conceiving) plainly involves more than, so to speak, thinking at the colour, or expending mental energy in its direction. It must involve some definite concept of whiteness (essence, theory of what whiteness is) which will *in fact* connect the colour with opacity, brilliance, achromaticity, etc. And with the concept of whiteness proposed in this chapter, the 'possible world' cannot be thought; it isn't a possible world at all. With no alternative concept before us, it remains a thoroughly indeterminate question whether we can think (conceive) or think of the possible world. (How do we tell when we've succeeded?) It will not do to say, pointing, 'I know what "white" means, here and in all possible worlds'. Of course

'white' means ———→ ◯. But what is ———→ ◯?

In the thirty years of colour TV the 'acceptable white' has changed substantially towards the *blue*. The older white now looks yellow. What is needed in order to determine whether it *is* yellow is not rootless intuitions about what we would say in this or that possible world, but rather a theory of what whiteness *is*.

(c) How do we determine the success of a reference in a possible world? Let us consider in outline Putnam's well-known proposal for determining the referent of a term like 'water' in possible worlds.[47] Take *this* – we point at a good sample of water in the actual world, and call it w^1. Now there is something which *explains* the typical properties or stereotype that a sample such as w^1 has – that it is clear, drinkable, freezes at a certain temperature under such and such conditions, and so forth. Call this something N for nature. Putnam's proposal is that w^2, a sample of something in a possible world called W_2, is water if it has N, whether or not w^2 has the same stereotype as w^1 does in W_1, the actual world. The key schema is '(For every world W) (For every *x* in W) (*x* is water if and only if *x* bears the relation 'same$_L$' to the entity referred to as 'this' in the actual world W_1). For w^1 and w^2 to stand in the theoretical relation 'same$_L$' to one another is for w^1 and w^2 to be the *same liquid* in the sense that they are both N (or both have N), whatever it is that explains the stereotype of w^1.

Putnam's scheme works beautifully for the problem of the identification of whiteness. In our world, the actual world, it is the high diffuse reflectance of white objects that explains all the typical properties of white – opacity, lightness, and so forth. By Putnam's method, if anything in any other world, actual or possible, has this same N, then it is whiteness. Perhaps there is a sense in which we can conceive a possible world in which whiteness has nothing to do with high reflectance – we *murmur* these words to ourselves – but as Putnam observes, even if it is conceivable that whiteness is not high diffuse reflectance, in the sense that we could invent imaginary experiences that would convince us, if they were real, that it was not high diffuse reflectance, still 'it isn't logically possible! Conceivability is no proof of logical possibility.'[48]

[47] Hilary Putnam, 'The Meaning of Meaning', *Mind, Language and Reality*, p. 231.
[48] Ibid., p. 233.

Then *do* we know a priori that we can or cannot imagine a transparent white object, either in this world or some other? *How*? (What happens when we attempt to imagine a colour? What fails to happen when the colour is impossible?) I think that we do know that a colour is conceivable or imaginable in the relevant strong sense only when the concepts involved have been analysed, scientifically or otherwise – that is, when we know or can say what those things are which they are the concepts of. The question why there cannot be a transparent white object is, so to speak, no stronger than the answer. A sincere belief that we can or cannot imagine or conceive non-Euclidean geometries, a trinitarian God who does not violate Leibniz's Law, or the construction of the heptagon, has no tendency to show either that these things are possible or that they are not. 'For many mathematical proofs do lead us to say that we *cannot* imagine something which we believed we could imagine'[49] and no doubt also that we can conceive something which we could not imagine – e.g. a line at infinity, the meeting of two parallel planes in a line with a function but no location.

(3) The third difficulty also concerns imagination. Has the question why we cannot imagine a transparent white object been answered? Why can't we form an *image* of something transparent and white? Why does the image of whiteness as it were block the image of transparency? The plain difficulty is that the refusal of the transparent colour to coalesce in the mind can have nothing to do with *physics*.

The answer I want to propose is that to show that the 'object' of the image (what the image is an image of) is impossible, in the sense that its description is logically contradictory, is to show that this object is unimaginable in the relevant Wittgensteinian sense. There is no need for a further explanation of why the image, or, better, the logically more basic imagining, is impossible. The round square is impossible, but there is no need for a psychological or phenomenological explanation of the hopelessness of the

[49] *Philosophical Investigations*, 517 (Oxford: Blackwell, 1976), p. 141e.

attempt to imagine or produce an image of it. What is impossible when it is impossible for us to imagine an object in the relevant sense is the *object*, not the imagining or the image.

And perhaps in the weaker sense neither the image nor the imagining are impossible. Consider *Remarks on Colour* I 43:

> A smooth white surface can reflect things. But what, then, if we made a mistake and that which appeared to be reflected in such a surface were *really* behind it and seen through it? Would the surface then be white and transparent?

So we can form an image of how something white and transparent would have to look. It would exactly parallel how something green and transparent would look. A transparent green glass could be mistaken for a glassy green surface reflecting a visible scene. But all this is not enough. Wittgenstein's description tells us how a transparent white surface would have to look or appear, if only it could exist. There is still the question why such an appearance couldn't actually *be* the appearance of something white and transparent. We know from Escher's drawings what certain impossible three-dimensional objects would have to look like, if only they could exist. This is not sufficient to show that they *are* possible, or, what is the same thing, that full descriptions of them are logically consistent under analysis.

2.11 The Concept of Transparency

In conclusion, I wish to examine a particular piece of psychological theorizing which illustrates the weakness of any psychological explanation of the puzzle propositions, and incidentally takes us back to the topic of transparency which was our starting-point. I have tried to show that there is a certain sense in which there is no gap between the way white appears and its physical nature.

> It must not be thought that ideas such as those of colour and pain are arbitrary and that between them and their causes is no relation or natural connection: it is not God's way to act in such an unruly and unreasoned fashion. I would say,

> rather, that there is a resemblance of a kind – not a perfect one which holds all the way through, but a resemblance in which one thing expresses another through some orderly relationship between them. Thus an ellipse, and even a parabola or a hyperbola, has some resemblance to the circle of which it is a projection on a plane, since there is a certain precise and natural relationship between what is projected and the projection which is made from it, with each point on the one corresponding through a certain relation with a point on the other. This is something which the Cartesians have overlooked ...[50]

The same points could be made about transparency, which is in a certain sense the opposite of whiteness. The piece of theorizing about transparency which I have chosen to examine, however, shows exactly the mistaken 'Cartesian' conception which Leibniz is attacking in this passage. In examining Metelli's theory, I wish to take particular note of the way in which the phenomena are divided into separate and distinct psychological and physical domains.

In 'The Perception of Transparency'[51] Metelli actually goes as far as to claim that 'transparent' has two meanings, one mental or perceptual and one physical. In the physical meaning, something is transparent if light can pass through it. On the other hand, 'If we mean to say we can see through something, then the meaning we intend to convey is perceptual.' Metelli gives examples of things which he says are transparent in the first sense, but not in the second, and vice versa. Air, he says, is physically transparent, but 'normally we do not speak of seeing through it'. Perhaps, though, we don't normally *speak* of seeing through air just because we always *do* through it. Nor, says Metelli, 'do we always perceive plate glass doors'. Plate glass and air are physically transparent but not perceptually transparent – we receive no impression of transparency from them. But surely we *do* see through plate glass, even if we don't say that we do. We don't 'perceive' air and plate

[50] Leibniz, *New Essays on Human Understanding*, trans. Peter Remnant and Jonathan Bennett (Cambridge: Cambridge University Press, 1981), p. 131.

[51] Fabio Metelli, 'The Perception of Transparency', *Scientific American*, April (1974), pp. 91–98.

glass because we see *through* them. Metelli's definition of perceptual transparency is that we perceive it when we see 'not only surfaces behind the transparent medium but also the transparent medium or object itself.' So air and plate glass become perceptually transparent 'when there is fog in the air or marks on the glass'. We have the impression of transparency, and Metelli's purpose is to isolate the phenomenological rules of its appearance. Here the question must be whether we see the perceptually transparent air or glass, or whether we see fog *in* the air and marks *on* the glass. What does perceptually transparent air *look* like, now that we can perceive it? If anything, it looks like *fog*. Nor does the *air* look transparent. What we see when we see dirty air is surely the dirt in the air, not physically transparent air transformed into perceptually transparent air. Moreover (here I follow David Katz), there is a sense in which we are aware of the transparency of the air around us. Differences in the clarity of the light are important in painting and commercial lighting, even though this does not have to do with one isolable sensation in the visual field.

Metelli's example of physical transparency without perceptual transparency is equally unsatisfactory. A transparent coloured card is placed on top of a black card. The transparent card 'no longer is perceived as being transparent; it appears to be opaque'. There are a number of difficulties here, including 'is perceived as being' (looks?), but the most important concerns the question *what* it is that looks opaque. The arrangement of the cards is such that it isn't possible to tell just by looking whether there are *two* layers of *one* card, the first transparent and the second opaque, or whether there are *two* cards, the transparent one on top of the opaque one. If the second, then it isn't true that the top card looks opaque, because we can't see the bottom card through it. Since the only difference between the two cases might be a thin film of transparent *glue*, it isn't possible to tell which it is without trying to separate the cards. But Metelli's subjects aren't allowed to do this. Perhaps if they saw the top card actually being placed on top of the bottom opaque card it would not be 'perceived as being' opaque. It would appear exactly as a transparent card on top of an opaque card should appear – transparent. The two cards together might be said to act together as *one*; but then we *should* say that they are opaque! *They* together are opaque.

Metelli goes on to say that Metzger has shown that mosaics of opaque card can 'give rise to' an *impression of transparency*. This 'give rise to' means that Metzger has prepared arrays which resemble three-dimensional scenes out of opaque cards. These arrays copy the patterns of colour, light and shade which are present when something actually is transparent, much like a photograph of something transparent, which is not itself transparent. But perhaps the impression is not really strong enough to make someone outside a restricted perceptual environment actually suppose that he is seeing something transparent. At best it could be said that Metzger has devised mosaics which under the appropriate conditions might be mistaken for the real thing. But because I make a cardboard tree which is good enough to be mistaken for a real tree, we should not conclude that we have an example of perceptual treeness without physical treeness.

'These two examples', Metelli continues, 'make it clear that physical transparency is neither a necessary nor a sufficient condition for the perception of transparency.' Physical transparency, as he puts it, cannot 'explain' perceptual transparency – just as physical treeness cannot explain perceptual treeness – I can see a perceptual tree (a cardboard tree) without a physical tree. To say that physical transparency is not necessary for the perception of transparency is only to say that something untransparent can be made to look more or less like something transparent. What looks as if it might be transparent isn't necessarily. But this 'looks as if it might have been transparent' *is not a kind of transparency*, and accordingly we should not reserve a special mental or perceptual meaning of 'transparent' for it. Ockham's Eraser (due to Katz) goes against the reservation 'Dictionary entries should not be multiplied unnecessarily.'

Total confusion breaks out with Metelli's deduction that since the light reaching the retina as input does not contain the 'specific information about the characteristics of the transparent layers through which the light has travelled and been filtered, perceptual transparency is not the result of "filtration"'. 'It is a new fact [*sic*] originating in the central nervous system as a result of the light stimuli acting upon the retinal cells.' The 'explanation' of the new fact is found in the central nervous system. In fact there is no such thing as perceptual transparency. There are only ambiguous arrangements of cards which are more or less like arrangements of cards which really *are* transparent.

Suppose I am fooled by a counterfeit £10 note. I have the perception of *money*. (Does 'money' have two senses, one for forgeries?) The same stimuli (money stimuli) hit my retina when I look at real money and when I look at a forgery. So the fact that what I see as real money is real money does not explain my seeing it *as* money. And I can see money – perceptual money – when I am not seeing physical money. What allows the perception of something as money that isn't money? Only *the perception of moneyness* which must 'arise' in the central nervous system – where else?

2.12 An Argument to Show that We Are Not Brain-in-a-Vats

Wittgenstein writes,[52]

> Explaining colour words by pointing to coloured pieces of paper does not touch the concept of transparency. It is this concept that stands in unlike relations to the various colour concepts.

But the distinction between physical and phenomenal transparency in Metelli's sense is false. Metelli hopes to 'explain' the word for phenomenal transparency by 'pointing to coloured pieces of paper' in certain arrangements. This approach must fail. Transparency does not lie in the image. It belongs, with whiteness, in the external world. This means that brain-in-a-vats would not be aware of it, since they are not aware of the external world. Since we are aware of it, it follows that we are not brain-in-a-vats.

[52] *Remarks on Colour*, III 189.

The various 'colours' do not all have the same connection with three-dimensional vision.[53]

> [What explains this] must be the connection between three-dimensionality, light and shadow.[54]

This does not mean, and nor would the point remain valid if it did, the connection between the *image* three-dimensionality, light and shadow. For one thing, there is no image of light in the required sense. The brain-in-a-vat is not exposed to *real* light, nor to three-dimensionality. This point can be summed up in a remark Wittgenstein makes at *Remarks on Colour* III 253 which highlights the problematic character of the concept of the visual image.

> What is the nature of a visual image that we would call the image of a coloured transparent medium?

The answer is that the *nature* of the visual image does not lie in the image.

2.13 Logic, Physics, Phenomenology and Psychology

If *W* & *Tr* is a contradiction in the truth-functional sense that it is false no matter what the truth-values of *W* and *Tr*, then the idea of a separate logic or grammar of colour is wrong. But *W* and *Tr are* contradictory under analysis. So *W* & *Tr* is in fact a false proposition, and therefore –(*W* & *Tr*) is a true proposition, and a necessary one as well. So the latter is not a manifestation of grammar in Wittgenstein's sense. And when we see *why* this is so, we can also see why –(*W* & *Tr*) is not a phenomenological proposition. The reasoning is as follows. Since *W* & *Tr* is contradictory, it must be capable of being analysed into certain simpler propositions, which I have called *r* and *m* (see Section 2.5). But (1) *r* and *m* are physical not phenomenological propositions, and since, as we saw in the previous section, *m* does not report a phenomenon, neither does *r*. We do not see light reflected into our eyes, as Wittgenstein points out; (2) if *W*

[53] Ibid., III 142.
[54] Ibid., III 144.

& *Tr* can be completely analysed, it is not a *sui generis* phenomenological proposition.

The view of colour advanced in this book differs from phenomenology, in the philosophical sense, over certain crucial truths about the puzzle propositions. They mark invariant features of colours because they are necessary truths, rather than being necessary, as in phenomenology, because they describe invariants. All this means, however, is that not being transparent is a necessary condition for being white, or

$$W \supset -Tr.$$

This proposition is derived from $-(W \& Tr)$, and one is necessary if the other is. Thus the essence of whiteness (or part of it) is reached not by a *Wesenschau*, or the intuition of an essence, but by logic and the definitions suggested by observation and experiment. So we need not treat colour phenomena – colours – *just* as phenomena, as 'pure phenomena' whose essential structure is to be intuited (unless this means *not* to organize them according to the wavelength theory).

> Physics wants to determine regularities; it does not set its sights on what is possible.
>
> For this reason physics does not yield a description of the structure of phenomenological states of affairs. In phenomenology it is always a matter of possibility, i.e. of sense, not of truth or falsity. Physics picks out certain points on the continuum, as it were, and uses them for a law-confirming series. It does not care about the rest.[55]

In order to 'care about the rest' we do not have to be phenomenologists. But what are these points which 'physics does not care about'? A physiological theory such as the Young–Helmholtz trichromatic theory of colour perception is only interested in points in colour space between the R, G and B primaries as functions of the primaries. Yellow, which is mixed from red and green, is regarded *as* a

[55] Ludwig Wittgenstein, *Ludwig Wittgenstein and the Vienna Circle* (Oxford: Blackwell, 1979), p. 63.

balanced red-green. What about the fact that yellow is a 'colour' in its own right, quite apart from the physiology and physics of its production? Physics doesn't care about the fact that there is such a colour as yellow. This would be *psychology*. Interestingly, Helmholtz claimed that he *saw* yellow as a red-green. The more responsible reaction is Ladd-Franklin's: 'Consciousness rebels against the trichromatic theory.'

Is there sense to the idea that there should be a unity between the deliverances of psychology – consciousness – and physics? If not, would consciousness be right to rebel against a physical theory? The unity of physics and psychology is achieved when physics becomes phenomenological. This is brought about in the theory proposed by the application of logic to the phenomenology. When phenomenology and logic are drawn together by the definitions of the colours, the result is a phenomenological physics. This is not surprising if we consider the various types of possible explanation which compete for the puzzle propositions.

LOGICAL GRAMMAR

PHYSICALISM PSYCHOLOGY OF SENSATIONS

PHENOMENOLOGY

With the definitions in place, logical grammar becomes, as we have seen, straight logic. Phenomenology, in the philosophical sense, becomes the setting out of the consequences of the definitions. The psychology of sensations becomes unnecessary, and the *content* of sensations – colours – goes into the definitions. And physicalism is replaced by a simplified phenomenological physics, which ignores the 'points' of physics, such as wavelength of reflected light.

LOGIC

PHYSICS → PHENOMENA

DEFINITIONS ← CONTENT

PHENOMENA → DEFINITIONS

3

Brown

> The solution of philosophical problems can be compared with a gift in a fairy tale: in the magic castle it appears enchanted and if you look at it outside in daylight it is nothing but an ordinary bit of iron (or something of the sort).
>
> WITTGENSTEIN

3.1 The Puzzles

Wittgenstein raises a number of puzzles about brown, but he does not seem even to attempt to answer them. They did not make their way into *Remarks on Colour* I. As with white, it seems to me that the puzzles about brown have everything to do with what the colour *is*, and that without this the puzzle propositions cannot be elucidated. The worth of the real definition of brown given in this chapter will be measured precisely by its ability to explain the asymmetrical properties of brown by implying the puzzle propositions. Here too Wittgenstein's way of drawing the line between logical or grammatical questions and scientific questions will be brought into question.

I shall also have things to say in this chapter about sensations, things which suggest that brown, the so-called phenomenal property, has nothing to do with mental 'states', whatever they may be.[1] In the proposed solutions to Wittgenstein's puzzles sensations are not mentioned because they are irrelevant.

[1] 'His mental state' is a concept in good order, but not 'his mental states'. If what philosophers have meant by 'state' in this connection is the instantiation of a property at a time and a location, e.g. John's believing that the world is round, John's having a headache, etc., matters become even more confused. It's *true*, say, that John believes that the world is round, but is this belief a *property* of his? What would this mean? The 'state' concept gives a false unity to a wildly heterogeneous set of phenomena.

Perhaps the solutions cannot even be stated in a theory in which brown is a property of an effect of the physical stimuli. But the explanations which are needed are about the brown (phenomenal brown, if we must call it that, but as opposed to physical brown, not noumenal brown) which we see, experience, are directly aware of, or whatever, and which Wittgenstein's puzzles are about. Part of the trouble is that there *are* no physical stimuli for brown. The stimulus must be light or electromagnetic radiation, and there is no brown light. Brown does not appear in the spectrum. I attack the confused view in which brown is regarded as a 'physiological' or 'induced' colour. My argument is intended to widen the concept of the physical by changing the physical stimulus for the colour. The answers to the puzzles are to be found, as with white, in a domain which is neither physiological nor psychological, but nor is it concerned with the quantity and colour of the light entering the eye. The visible impurity of brown is explained neither by the physics of the stimulus nor by neural coding: it lies in the concept of the colour.

The following eight remarks or groups of remarks are the explananda whose explanans will be (if all goes well) a statement of what brown is. These remarks are all taken from *Remarks on Colour* III.

(i) 46. Among the colours: Kinship and Contrast. (And that is logic.)

(ii) 47. What does it mean to say, 'Brown is akin to yellow'?

(iii) 48. Does it mean that the task of choosing a somewhat brownish yellow would be readily understood? (Or a somewhat more yellowish brown?)

(iv) 49. The coloured intermediary between two colours.

(v) 60. Why don't we speak of a 'pure' brown? Is the reason merely the position of brown with respect to the other 'pure' colours, its relationship to them all? – Brown is, above all, a surface colour, i.e. there is no such thing as a *clear* brown, but only a muddy one. Also: brown contains black – (?) –

How would a person have to behave in order for us to say of him that he knows a *pure*, *primary* brown?

(vi) 62. What does 'Brown contains black' mean? There are more or less blackish browns. Is there one which isn't blackish at all? There certainly isn't one which isn't *yellowish* at all. (The MS may contain a question mark here. Ed.)

(vii) 65. 'Brown light'. Suppose someone were to suggest that a traffic light be *brown*.

(viii) 215. Why is there no brown nor grey light? Is there no white light either? A luminous body can appear white but neither brown nor grey.

3.2 A Definition of Brown

It happens that there is a single decisive explanans for all of these propositions, insofar as they concern brown. There is a well-known experiment which illustrates the physiological or psychological effect of so-called 'induction'. A uniform expanse of brown is viewed through a narrow tube lined with black velvet. After a very short time the brown disappears and is replaced by a good yellow, the exact hue depending on the colour of the original brown. A surface will turn brown if it is yellow and darkened relative to the surroundings. The surrounding field in the tube is very dark, and when this replaces the bright daylight around the brown, we get the 'non-induced' mode of the colour – yellow. The brown is supposedly *induced* into the yellow. I shall call this experiment 'A'. R.M. Boynton writes,

> A chocolate bar and an orange have approximately the same chromaticity, and yet their colours are entirely different because their reflectances vary so greatly. For the chocolate, the sensation of brown arises *de novo* by induction from the surrounding field.[2]

[2] R.M. Boynton, 'Colour, Hue and Wavelength', in E.C. Carterette and M.P. Friedman (eds.), *Handbook of Perception* V (New York: Academic Press, 1975). p. 316. Chromaticity is a function of hue and saturation. Cf. R.L. Gregory, *Eye and Brain* (London: Weidenfeld, 1977), p. 127: 'Brown normally requires contrast, pattern and preferably interpretation of areas of light as surfaces of objects

Later on I shall question this way of describing the matter. The facts as described by Boynton are not obviously relevant to Wittgenstein's puzzles. Consider, however, an experiment I shall call 'B', which illustrates the same effect. Projector 1 is fitted with a chrome yellow filter and an opaque slide with a clear round centre, and projector 2 with a slide having an opaque black centre but otherwise clear and transparent. When projector 1 alone is switched on, we get a chrome yellow spot on the screen. When projector 2 alone is switched on, we get a black spot on a white field. When the two projectors are switched on together and the two spots brought into register, the surroundings of the yellow patch become brighter than the yellow patch – it now appears as a darkening – and it turns a good chestnut brown. This experiment can be made with all the main hues simultaneously. Red-yellows turn brown, and green turns olive. 'Thus red, yellow and green of low luminosity are red-brown, brown and olive respectively' Helmholtz writes in the *Physiological Optics*.[3] Notice this use of 'are': 'brown is a yellow of low luminosity' must be an identity statement. But Helmholtz also says, inconsistently, 'The writer has succeeded in making the homogeneous golden yellow of the spectrum look brown' by 'illuminating a little rectangular area with this yellow light on an illuminated white screen; while at the same time a neighbouring portion of the screen was illuminated by brighter white light.'[4] This phrasing illustrates Helmholtz's well-known but incorrect belief that contrast and related effects can be attributed to errors of judgement or unconscious inference. The truth is that a yellow coloured *object* can look brown, but the *colour* yellow cannot, for then it

---

(such as wood) before it is seen, and yet in normal life it is one of the most common colours.' What is the significance of this 'and yet'? Why should it be surprising that a colour which requires contrast, etc., should be a common colour? I also have a doubt as to whether there are any experiments which show the need for pattern and interpretation in the creation of brown; Gregory does not cite any. My doubt is based on the fact that experiment B produces brown only on the basis of darkening or contrast.

[3] Vol. II, ed. J.P.C. Southall (New York: Dover, 1962), p. 130.

[4] Ibid., p. 131.

what?) would *be* brown. 'In such a situation it either vithdraws or ceases to exist.'[5]

It is interesting to see how, in B, the union of the two pots when they are taken out of register is still brown. The ffect amusingly resembles colour mixing. It is as if the econd projector were fitted with a *black* filter! When the ellow spot is projected onto the white screen, the colour of the spot appears as a light. When we project the dark pot onto the yellow one, it appears that the new spot is ltering the prevailing yellow-white illumination. B simuates what actually does happen when a dark area modifies he illumination. The brown spot appears because the illunination is effectively darkened at that point. What maters is whether the colour datum appears as an unmodified ight or as a changing of the light – a shadow. A printed ellow spot surrounded by greater brightness, as in B, will lso turn brown. How are these effects related to the mixure of black, red, and yellow pigments? Are they additive r subtractive processes? Or must we regard them as lifferent from either?[6]

My suggestion is that we should say, skipping any refernce to sensations or induction, that: *the colour brown is kind of darkened yellow*. This follows Goethe's elegant rawing together of 'conditions of appearance', description, nd essence of a colour. 'Darkened' in the formula is meant o distinguish the contrast conditions achieved by the objective' or non-physiological version B of the Helmholtz xperiment from (a) dark yellow and (b) yellow of low

[5] Plato, *Phaedo*, 102e.

[6] Michael Wilson has suggested that the received distinction between additive nd subtractive colour mixing can be replaced by the more fundamental distincon between processes of lightening (additive) and processes of darkening (subactive). From this point of view, the brown effects described above do not need be regarded as 'subjective', 'induced' or 'physiological', except in the sense in hich all colour perception is physiological. They are rather dramatic illustrations f the role played by contrast and adaptation, lightening and darkening, in the erception of *all* colours, which are concealed in the usual way of describing dditive and subtractive mixing. It is *not* obvious that the processes involved in ixing green from blue and yellow pigments are fundamentally dissimilar from e mixing of brown from black and yellow pigments; and this is the same effect A and B.

luminance.[7] The significant fact about brown objects is th they are yellows which darken the light more than u modified yellows can.

3.3 The Puzzles Solved

In what follows, this definition of brown is deployed solve Wittgenstein's puzzles, which I shall take in the abov order. The puzzle propositions are logical consequences the definition of brown, but I shall be making no clain about the meaning of 'brown' or about analyticity. Rathe the claim will be for the 'being of brown, whereby it is wh it is': this is a theory of the essence of the referent 'brown'. I shall not discuss (i) 'Among the colours: Kinsh and Contrast. (And that is logic.)', because I hope th what sense it does have will emerge from the discussion (ii)–(viii).

(ii) 'What does it mean to say, "Brown is akin to yellow" Brown is akin to yellow in the strong sense that it is a *kin* of yellow – the darkened kind. It is akin to yellow in hu but not in saturation and brightness. The kinship is simil to the kinship between pink and red. Pink is a lightene or whitened red, and there are therefore no pin which are not reddish.[8] This clears up another apparent

[7] For a criticism of this kind of definition, see C. James Bartleson, 'Brow *Color Research and Application* **1** No. 4 (1976), pp. 188 and 190. 'Brown is n simply "a dark orange", for example. Brownness increases in inverse proporti to lightness while orange is not proportionately related to lightness. Orangene increases in direct proportion to strength while brown is not proportionate related to strength.' (1) Bartleson does not distinguish a dark orange (dark for orange) from a darkened orange. (2) The fact that what is true of orange is n true of brown does not show that brown is not a *kind* of orange, or yellow. M claim is not that brown is identical with orange; but this is the propositi Bartleson appears to be arguing against. Cf, 'Ice is a kind of water – the froz kind: but ice is solid and water is liquid'. One could say, 'Ice is water'. between yellow and orange, my view is that it is the yellow in the orange which responsible for the production of brown from orange. On the other hand, the be browns are definitely from the warmer or redder yellows.

[8] This settles the fate of that amphibolous 'sentence of ill-fame, "Pink is mo like red than black"' (J.L. Austin, 'The Meaning of a Word', in *Philosophic Papers* (Oxford: Oxford University Press), p. 66), in both meanings ('Pink more like red than black is like red' and 'Pink is more like red than pink is li black' (cf. 'Oxford is nearer to London than Cambridge'). The remaining que tion is why some reds differ from others in a certain ill-defined respect resembli

unrelated puzzle from Wittgenstein: why is there no such thing as a blackish yellow? Why isn't a dark yellow at all blackish?[9] Here we need to know that yellow is a relatively bright colour at maximum saturation, and that a dark yellow therefore has roughly the same brightness as a relatively light blue. 'Dark' in 'dark yellow' is an attributive term. It means 'dark for a yellow'. Now if brown really is a kind of yellow, we have the explanation of what has happened to the missing blackish yellows, which are darker than the darkest yellow. *They are browns*. Browns can be blackish. The interesting question is why yellow is unable to tolerate darkening or dirtying, as Goethe stressed, and why it undergoes something resembling a change of hue if it is made to appear as a darkening of the light. What is the relationship of yellow to the light?

(iii) 'Does it mean that the task of choosing a somewhat yellowish brown would be readily understood? (Or a somewhat more yellowish brown?).' This seems back to front. 'Brown is akin to yellow' certainly doesn't *mean* that the task would be more readily understood, except in the sense that it implies that it would be or could be. If it did mean that, in the sense of being synonymous with it, we would be stuck with the question ('Wittgenstein's Spade') *why* the task would be or could be readily understood, and the comparable task of choosing a somewhat *bluish* brown would not be or could not be understood. (Why can't there be a bluish brown?) The task of choosing a yellowish brown would be understood because there are, between yellow and brown, a number of intermediate hue steps, all of them yellow or brown or undecidedly either. The task would be understood because the intermediate hues are there, because of the 'kinship', and not the other way round. The intermediate hues are there because brown is, so to speak, yellow under new conditions: because yellow turns brown or turns *into* brown when it is darkened.

---

hue, and why this should earn them their own colour name when there is no comparable name for e.g. the whitish blues.

[9] *Remarks on Colour*, III 106.

'Why don't we speak of a pure brown?' cannot be the right way of putting (v). We learn nothing extra from the linguistic formulation of the question, as we learn nothing from the formulation of *Remarks on Colour* I 42, 'We speak of a "dark red light", but not of a "black-red light".' Whether we speak or don't speak of these things cannot be the important question. Clearly 'dark' in 'dark red light' signifies the kind of red, not the kind of light, whereas 'black' in 'black-red light' signifies a component of the illumination (note the hyphen), and qualifies the light. 'Black-red' would function like 'orange-yellow' or 'blue-green'. In the more perspicuous non-linguistic formulation which Wittgenstein also adopts, e.g. at *Remarks on Colour* III 24, the question will be, 'Why can't there be a black-red light?' Now if there can be an X-Y light, there can be an X light, for X-Y can be made more X and less Y. So if there can be a black-red light there can be a black light . . . a black *light*? And now the question is hard but clear. What *is* black? Why aren't there any light blacks? Why is black the darkest colour? And what is a light?

(iv) is a curious remark: 'The coloured intermediary between two colours.' It is not a sentence. What is it doing where Wittgenstein puts it? Perhaps the point is that it connects the idea of two 'kin' colours at III 46 and 47 with the continuous hue sequence latent in III 48. Brown and yellow would be 'kin' because of the fact that no other colour lies between them in colour space. There is a *direct route* from yellow to brown, but none from yellow or brown to blue. But now the question reappears. *Why* can't another colour intervene? Why *can't* it? Why is the geometry of colour space what it is? This is not at all like the empirical question why there is no direct rail route from East Grinstead to Oxford (London = grey). Colour space is Leibnizian. The place occupiers determine the geometry of the space. So the question is what blue, brown and yellow would have to *be* in order for the direct route to be conceivable. What would the colour black have to be in order for a dazzling black light to be conceivable? If indeed brown is a darkened yellow, there can be no *hue* difference between brown and yellow, and nothing of a different hue can come between them. There can no more be a coloured intermediary between yellow and brown than there can be one between light blue and dark blue.

(v) 'Why don't we speak of a "pure" brown?' I shall assume that by 'pure', Wittgenstein here means 'saturated' and not 'unitary', although he knows a *pure, primary* brown suggests some confusion. Does he mean 'pure, in the sense of primary (unitary)' or 'pure, and a primary as well'?[10] Generally in *Remarks on Colour* 'primary' means 'unitary,' and not 'additive primary' or 'reference colour' as in colour TV (red, green, violet) or colour printing (yellow, cyan blue, magenta) – most of these are non-unitary.[11] So I shall take Wittgenstein's question to be the question why there cannot be a saturated brown. The absence of brown in the spectrum ('monochromatic' brown) does not explain the unsaturatedness of brown, i.e. the fact that it has no place on the hue circuit. For there is no such thing as monochromatic purple or spectral purple, and yet purple has a place on the hue circuit.

If brown is what I have said it is, then there is no *light* which will produce it rather than yellow. There is light of no wavelength save the yellow at 590 nm. or whatever which will produce the colour under the given conditions. There cannot be a saturated brown because saturated brown would be yellow. There can of course be a *rich* brown or a *mid*-brown, but that is not at all the same thing.

[10] It is worth noting that it is not correct to say that there are different *senses* of 'primary' used for additive, subtractive and 'psychological' primaries, as the unitary hues are sometimes called. Primaries are primary with respect to (i) a particular purpose, e.g. printing or primary school painting, and so (ii) to a particular method, e.g. light mixing, rotary mixing, (iii) to a particular desired gamut of colours, or to an acceptable number of resultant hues: and so (iv) the primaries must be as few as possible. A primary is a colour or hue which will, with other primaries, yield a gamut satisfactory for some purpose. Brown could be *used* as a primary, as e.g. cyan or violet is. But it would not be a very *good* primary. Wittgenstein may mean that there is no unitary brown, or central paradigm brown, or 'true' brown, even though brown is counted as a colour. This is what one would expect if brown, apart from being a colour in its own right, is a kind of another colour. There is no central or primary light blue.

[11] For a discussion of the unitariness of brown, see Bartleson, 'Brown'.

Why is there no clear brown? We can either say that there cannot be a clear brown, because a good clear spectral brown would be yellow, or we can say that there is a clear brown: when brown is clear it becomes yellow, when yellow becomes unclear through failing to transmit light and absorbing it instead, it turns brown – muddy. This is not obviously because brown is a surface colour. The 'i.e.' in 'Brown is, above all, a surface colour, i.e. there is no such thing as a *clear* brown, but only a muddy one' is puzzling. White is above all a surface colour, but this doesn't mean it has to be muddy. Why should surface colours have to be muddy? (Would mud be muddy if it weren't brown? Would green mud be slime, not mud? Would pure white mud be muddy? Here colour *constitutes* the object.)

(vi) 'Brown contains black.' This is not quite convincing. Light browns, tending to amber, café au lait or fawn, do not appear to be blackish. Perhaps what Wittgenstein means is that brown pigment is mixed from black pigment and some other, and in this literal sense contains black. The effect of the black is to darken the mixture, and the limit of this darkening is black.

(vii) and (viii) concern the impossibility of brown light and the brown traffic light. We have seen that brown things must be relatively dark relative to their surroundings and to the illumination. Brown is a kind of shadowing. (A brown surface must absorb most of the available light and so it cannot also transmit it, or be transparent or clear.) So brown can never stand out from its surroundings; it can never appear blazing or brilliant, although it can be glossy. A brown traffic light would be lower than the surroundings in brightness. It would not shine. In dark surroundings, at night, the traffic light would be brighter than its surroundings, and *it would turn yellow*. Brown light is yellow. (This is not, of course, to say that there are not light browns. Light browns are not brown lights.) Or we could say that there cannot be brown light because brown darkens rather than lightens. A brown light would have to darken what it illuminated, and so it would be no kind of light.

3.4 Criticism of Psychology

Discussion of the genesis of brown as described by Boynton was postponed. It is now time to take up some of the difficult questions which his description raises. His idea is that a chocolate and an orange have the same chromaticity, but their colours differ because they differ so greatly in reflectance. 'For the chocolate, the sensation of brown arises *de novo* from the surrounding field.'

The first question to be asked is what sensations Boynton thinks *don't* arise *de novo*, and in what sense. The function of the concept of induction here is entirely metaphorical. Colours can appear in places in the visual field without any physical change in that place, but only a relational one to the illumination. The colour is induced into the place, but in some undefined sense it isn't really there. But then what is? The appearance of *all* colours depends on the relation of the eye to the light and to darkening or spectral absorption, to contrast, edges and adaptation. An absolutely stabilized image will rapidly lose colour completely.[12] Are all colours induced? So what is really there? Only a realist who knows what is there (electromagnetic radiation) and what arises *de novo* (brown – but what about grey, black and white?) can talk as Boynton does. We should not say that grey is induced into a white field by contrast, but rather that it is a colour whose appearance depends on a certain relationship or apparent relationship to the illumination. Incidentally, it is not, as Boynton says, the difference in the reflectance of the chocolate and the orange which is responsible for their different colours, but rather the difference between: the difference between the reflectance of the orange and the reflectance of its surrounding field; and the difference between the reflectance of the chocolate and its surrounding field, which is a cross ratio or ratio of ratios, or between 'dark' and 'darkened' in the senses outlined above. The

[12] A stabilized image is one which is made to move with the eye, so that retinal images do not change with time at each point on the retina. So there is no relative rapid eye movement, and the eye cannot make comparisons and contrasts.

important difference is between the colour-surround differences, not between the absolute reflectance values.

Fig. III is therefore misleading, in that it represents brown as a simple ratio rather than as a cross ratio. The relation to the illumination cannot be included in observations of the kind recorded in Fig. III. Fig. III could, for all the information given, represent a dull yellow object. A similar amendment should be made to all the definitions proposed in this book. The point is that a brown surface alters neither its reflectance nor its luminance when viewed through Helmholtz's tube in A, and the yellow spot in B is physically unaltered by the superposition of the black spot. The spot does however alter its apparent relative reflectance, and this is what affects its colour.[13] And the suppressed major premise of Boynton's argument is surely false: 'Their colours vary because their reflectances vary so greatly.' Two blue things of the same hue with different reflectances are not different colours. The change occurs only in the case of brown and yellow.

What seems to have gone wrong with Boynton's thinking is something like this. He accepts the general theory in which what *really* determines colour is the frequency of the light entering the eye from the sample. But this doesn't work for brown. There is no light which will produce brown. So Boynton introduces an extra story for brown, instead of doubting the general story about the other colours. Brown must be physiologically generated and arise '*de novo* by induction from the surrounding field'. Boynton has forgotten that he is also working with what is essentially Newton's division of the visual system in which all colours inhere in sensations stirred up by the Rays, not in the Rays themselves. So it must be the *sensation* of brown which arises *de novo*. But *de novo* relative to what? The surrounding sensations? *They* all arise (*de novo*) together. It is

[13] James Clerk Maxwell, 'Theory of Compound Colours', in David L. MacAdam (ed.), *Sources of Color Science*, p. 72, makes the mistake: 'Brown colours, which at first sight seem different, are merely red, orange or yellow of low luminance, more or less diluted with white.' A yellow of low luminance is a dim yellow, not a brown.

Fig. III

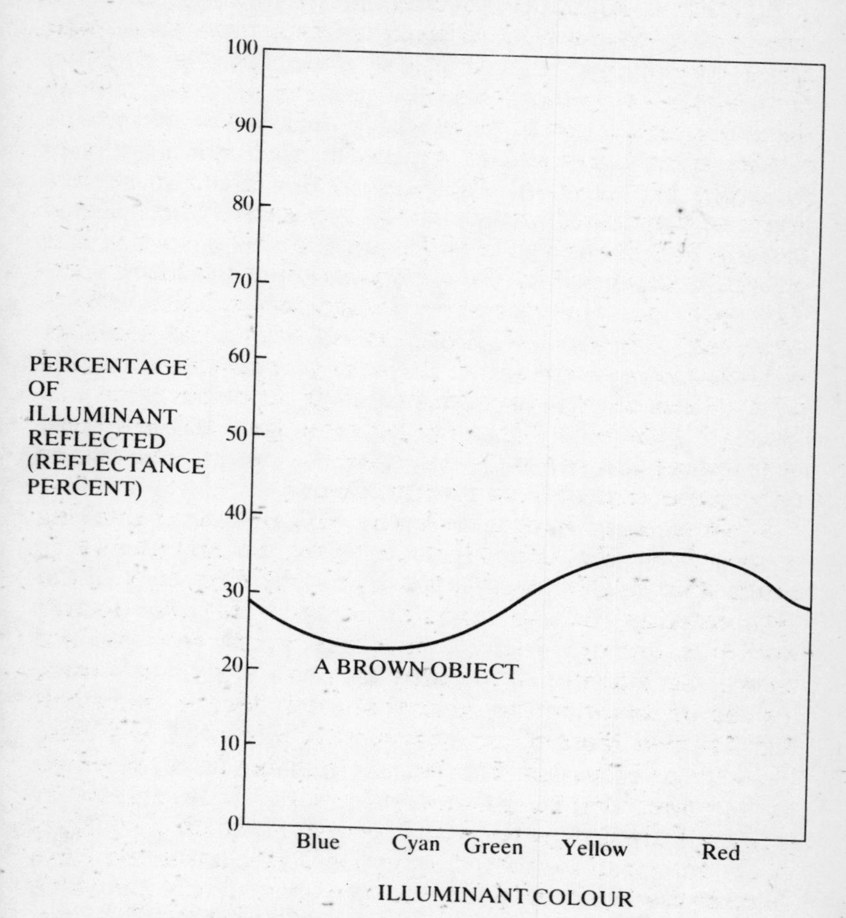

SPECTROPHOTOMETRIC CURVE FOR A BROWN OBJECT

not as if in B we end up with a sensation of yellow and brown induced into it. There *is* no sensation of yellow. It has ceased to exist. Or does brown arise *de novo* relative to the surrounding frequencies of light? But relative to *these* items, all colours arise *de novo* together. The Rays are not coloured. All that really needs to be said is that brown cannot be produced by *light*. The problem is how to square this with the general theory in which the colours of objects are determined by the spectral composition of reflected light. Brown could not be handled by what Land calls the 'classical tradition' (Newton, Young, Helmholtz, Maxwell) because it 'dealt with spots of *light*, and particularly pairs of spots, trying to match one to another.'[14]

Could Boynton's description of the genesis of brown be used to solve Wittgenstein's puzzles? Wittgenstein denied that a psychological account of the origin of the sensations would answer the questions as he meant them. There would be the further question why the sensations behaved as they did, and this would lead ultimately to a physiological explanation. Here we would not find the *logical* impossibilities which are needed. A different physiology could be conceived which *would* allow a pure brown, and so the impossible would after all be possible.[15] The same kind of

[14] Edwin H. Land, 'Experiments in Color Vision', in R.C. Teevan and R.C. Birney (eds), *Color Vision* (New Jersey: Van Nostrand, 1961), p. 165. There may after all be a way in which brown could be said to have a place in the spectrum. Michael Wilson has demonstrated how brown can be produced from an *open spectrum* (one in which the yellow/red and blue/violet edge spectra are far enough apart not to produce green). If the sharp boundary which produces the yellow/red band is replaced by a soft or graduated edge, the yellow/red turns brown/reddish brown (see M.H. Wilson and R.W. Brocklebank, 'Goethe's Colour Experiments', *Year Book of the Physical Society* (London: Physical Society, 1958), p. 165). Nonspectral purples can also be produced by the superposition of these graduated edge spectra. If we accept Goethe's account of the origin of the spectrum, we can say that brown has a place in it, because the concept of the spectrum has been widened. In the narrower sense, of course, it is just a fact that brown does not appear in the spectrum. Wilson regards his beautiful experiment as a counterexample to my definition of brown. I regard it as a striking confirmation. The dispersion of the unsharp edges through the prism gives a relative darkening of the colours of the edge spectrum.

[15] Cf. Wittgenstein's splendid remarks to Schlick about what should be said to a philosopher who believes in the synthetic a priori: 'Anti-Husserl', in *Ludwig Wittgenstein and the Vienna Circle*, ed. Brian McGuinness (Oxford: Blackwell, 1976), p. 67.

difficulty attacks physiological theories, such as the Hering scheme of opponent processes or Ladd-Franklin's schema of the evolution of the colour sense, which purport to explain the impossibility of reddish green and bluish yellow. Nor is there any need to 'psychologize' the explanation given above by saying that the sensation of brown is a sensation of a darkened yellow. The sensation terms as it were cancel out, and we are left with the explanatory element that brown is a darkened yellow. The solution to Wittgenstein's puzzle remains conceptual, not psychological. But this is not to say it is not scientific.

3.5 Why Is Brown a Colour?

There is a residual problem, on current theory a psychological problem, about the status of brown as a colour in the same category as red, green, blue and yellow. When B is performed with red, blue and green, the resulting maroons, navies and dark greens seem to retain their parent hue in a way in which brown does not. This must be due to the special position of yellow high on the brightness scale. In order to get the problem into proper perspective we would need an account of the significance of large changes in relative brightness for hue, as well as a proper theory of the conditions for unitariness of hue. Of the dark colours created by experiments of the same type as B, Boynton writes 'brown is certainly the most surprising because it ceases almost entirely to resemble the original bright colour.'[16] To me brown does *not* cease to resemble the yellow in hue. And of course in brightness there is no resemblance, but there must be a kinship of hue; or else all our colour scaling systems which assign brown a hue between yellow and red are wrong. The real issue is the inability of yellow to maintain its identity, as it were, when it is darkened. There is clearly something in Goethe's account of the relation between the quality of yellow and light, and also of blue and darkness. We need to know what it is, and this will deepen our understanding of the

[16] Boynton, 'Colour, Hue and Wavelength', p. 316.

peculiar character of brown. For the necessity of the proposition that saturated yellow is brighter than saturated blue suggests that an explanation of this fact in terms of the relative visibility curve and retinal sensitivity would be back to front.

3.6 Concluding Remarks

It is a pity that Wittgenstein chose to set up road blocks between the logic of colour concepts projected in *Remarks on Colour* and the physics, physiology and psychology of colour. 'We do not want to establish a theory of colour (neither a physiological one nor a psychological one), but rather the logic of colour concepts. And this establishes what people have often unjustly expected of a theory.'[17] The fact that we cannot conceive a certain impossible colour may not 'belong', as Wittgenstein puts it, to the physiology, physics and psychology of colour,[18] but our powers of conceiving have unquestionably been sharpened by advances made in these disciplines.

Wittgenstein also says[19] that 'phenomenological analysis (as e.g. Goethe would have it) is analysis of concepts and can neither agree with nor contradict physics.' This remark is perhaps inspired by the letter from Runge to Goethe, 3 July 1806, reproduced at the end of the Didaktischer Teil of the *Farbenlehre*, in which Runge comments on the theory, 'auch wird diese Ansicht den physikalischen Versuchen, etwas Vollständiges Uber die Farben zu erfahren, weder widersprechen, noch sie unnötig machen'. (This letter is also the origin of *Remarks on Colour*, I 21, III 94, III 105). But in Wittgenstein's hands the point is narrower and sharper. He seems to have believed that the sciences could contribute nothing to our understanding of the 'internal logic' of propositions, merely because they issue in *facts*. I do not believe that the incorporation of empirical facts prevents the explanation of the impossibility of pure brown

[17] *Remarks on Colour*, I 22.
[18] Ibid., I 40.
[19] Ibid., II 16.

and the other asymmetrical properties of brown from issuing in necessary truths. The truth expressed in a real definition must be an empirical one, whatever else it may be, but with the truth fixed logical consequences will follow. There is nothing here for the tender-hearted phenomenologist or philosopher of language to fear.

But he is right to fear the muddled Lockean and Cartesian 'mirror' image of the mind which dominates the physiology and psychology of colour today, and he is right to fear reductive physicalism as proposed by J.J.C. Smart and David Armstrong. These theories would replace colours by things which have only the most dubious claim to resemble them or model their properties – blobs of consciousness and changes in electromagnetic potential – and whose only virtue is that they are regularly found in textbooks of elementary physics and psychology. If the concept of brown is what I have said it is, then the truth is that this colour at least could not be a 'phenomenal property', or inhere in a mental state or whatever. Nor is it anything physical in the physicalists' sense. It is not logically simple, and it should not be confused with 'the experience of brown' – brown with a coating of apperceptive phosphorescence, as it were – whose existence I doubt. I have no experience of *seeing* brown, or if I do I cannot distinguish it from the experience of *seeing* which I have when I see any other colour. What is different is *what* I see. Nor can I on introspection detect anything which could really be called the 'experience of brown'. I find that I am not at all sure what this phrase means, or could mean.

Wittgenstein's great merit was that he saw with complete clarity what colours are not, and what they *could* not be. His puzzles provide a kind of test which any satisfactory theory of colour must pass. But Wittgenstein denied himself the resources to make any headway with the question of what colours are. What stood in his way, if what has been said so far is along the right lines, were: (1) his temperamental mistrust of the notions of the character and essence of a colour, and of any determinate answer to the 'What is it?' question, expressed in the thesis of the indeterminateness of colour and the claim of the *Philosophical*

Investigations that essence is revealed by grammar;[20] and (2) his doubts about scientific explanation. If brown and white are anything to go by, grammar flows from essence, and essence is revealed by science or any other activity which contributes to the answering of the 'What is it?' question.

[20] *Philosophical Investigations*, 371.

4

Grey

Yon grey is not the mornings eye.

SHAKESPEARE

4.1 The Puzzles

Like brown and white, grey can be regarded as a kind of darkening in the sense outlined above. A number of its asymmetrical or puzzling properties parallel those of brown, and similar puzzle questions result. From the *Remarks on Colour*:

(i) I 34. There is the glow of red-hot and of white-hot; but what would brown-hot and grey-hot look like? Why can't we conceive of these as a lower degree of white-hot?

(ii) I 36. Whatever *looks* luminous does not look grey. Everything grey *looks* as though it is being illuminated.

(iii) I 40. For the fact that we cannot conceive of something 'glowing grey' belongs neither to the physics nor to the psychology of colour.

(iv) I 41. I am told that a substance burns with a grey flame. I don't know the colours of the flames of all substances; so why shouldn't that be possible?

(v) III 80. What makes grey a neutral colour? Is it something physiological or something logical?

(vi) III 83. Grey is between two extremes (black and white), and can take on the hue of any colour.

(vii) III 96. Grey is not poorly illuminated white.

4.2 The Definition of a Grey Area

I shall answer the puzzle questions suggested or stated in these remarks, as before, with a definition, in this case of a *grey area*. Like white and black ones, grey areas, or surfaces, have approximately the same reflectance for light of all colours. They do not change the colour of the illumination, but they differ from white areas in that they do reduce the quantity. The grey card whose spectrophotometric curve is shown in Fig. IV reflects 20 per cent of any illuminating light. This means that it is relatively dull under any illumination. As with white, black, and brown, it is not the absolute value of the reflectance that determines the colour. An illuminated pale grey surrounded by darkness becomes whiter, and the same sample surrounded by something brighter becomes black. A bright light shone onto a black spot suspended in mid air against a normally illuminated background will turn it grey, and even white. The definition accommodates these facts, in that their description will be of situations in which the quantity but not the quality of the illuminant is apparently unchanged, i.e. ones which seem to darken white or colourless light but in which its colour is not altered. This definition captures the achromaticity or neutrality of grey.

4.3 The Puzzles Solved

(i) Why is there no grey-hot, and why couldn't we conceive it – as a comparison of Figs. I and IV might appear to suggest – as a lower degree of white-hot? A glow is an incandescence, a *bright* emission of light. A grey area is a relatively *dark* area. An area that was both glowing and grey would be both relatively dark and relatively bright. (This is not strictly a contradiction, since 'bright' and 'dark' are contrary rather than contradictory terms, but I assume that an adequate analysis of 'bright' and 'dark' would produce one.) If a torch is shone into coals which have been glowing a warm orange in the dark, the result is disappointing. The coals turn grey and cease to glow. They have become less bright than their surroundings. 'A colour "*shines*" in its surroundings. (Just as eyes smile in a face.) A "blackish" colour – e.g. grey – doesn't "*shine*"'.[1] Grey participates in the *darkness* of black. Something white and

[1] *Remarks on Colour*, I 55.

Fig. IV

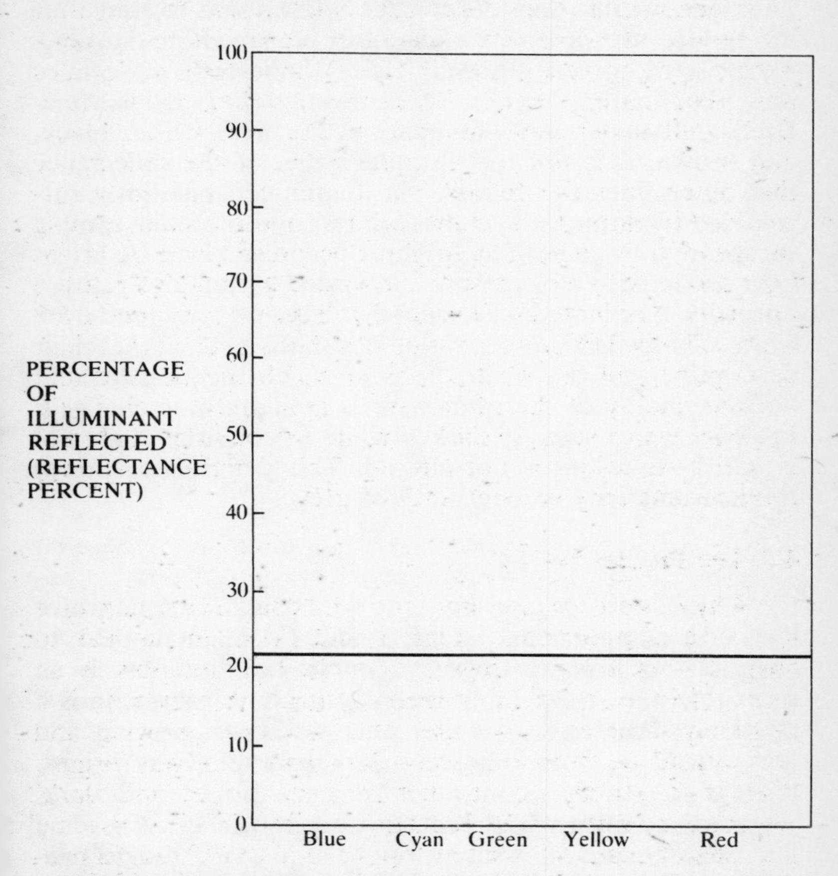

SPECTROPHOTOMETRIC CURVE FOR A PERFECTLY GREY CARD

hot can glow, but since something grey or black cannot glow, it cannot glow grey-hot or black-hot.

Why do we suppose that we could 'describe the colour impression of a surface area by specifying the position of the numerous small coloured patches within this area'?[2] The colour of many objects is specified in the opposite way, by the overall impression or *main* colour impression, e.g. botanical specimens in 'a good Northern light'. 'It is only *to be expected* that we will find adjectives (as, for example, "iridescent") which are colour characteristics of an extended area or of a small expanse in a particular surrounding ("shimmering", "glittering", "gleaming", "luminous").'[3] It is only to be expected ... What is perhaps surprising is that 'grey' falls into this class of terms. But then if the argument of Chapters 2 and 3 is correct, so do 'white' and 'brown'. This is not some vague sub-Hegelian point about the dependence of everything on everything around it, but a perfectly precise fact about the genesis and nature of colour phenomena.

(ii) 'There is no such thing as a luminous grey', Wittgenstein declares at III 81. And at I 37 he says 'What we see as luminous we do not see as grey.' If by 'luminous' Wittgenstein means 'shining bright', then the explanation is that since grey is a relatively dark colour, it cannot be luminous. On the other hand we can describe the suffused radiance of the sky as luminousness, though perhaps it could not be described simply as *radiant* except in the physicist's sense. Wittgenstein seems to have understood by 'luminous' something which is itself a light *source*. 'But the sky which illuminates everything that we see *can* be grey! And how do I know merely by its appearance that it isn't itself luminous?'[4] Perhaps it isn't correct to say that *the sky* illuminates everything that we see. It is surely the sun which illumines, and its light is diffused, not through the sky conceived as a hard material hemisphere with a definite

[2] Ibid., III 58.
[3] Ibid., III 66.
[4] Ibid., III 219.

radius, but through the cloud cover. (Is the *day* then grey? What is a *day*?) So this question is really (viii), why there cannot be a grey light. Why cannot grey be the colour of a light source, a shining or a glowing?

(iii) The fact that the impossibility of glowing grey derives from the concept, definition or theorita of the colour means that I am actually in agreement with Wittgenstein on the crucial *Remarks on Colour* I 40: 'For the fact that we cannot conceive of something 'glowing grey' belongs neither to the physics nor to the psychology of colour.' Wittgenstein would not have agreed to the alignment made here of concept, scientific theorita or definition, and essence. For him, a concept plays a role only within a language game. So for him the inseparability of the concept from the elementary phenomenological physics or ecological optics in Gibson's sense would represent a puzzle. But as we saw earlier, he does declare, at *Remarks on Colour*, III 180: 'We are not concerned with the facts of physics except insofar as they determine the laws governing how things appear.'

(iv) Why is a grey flame impossible? A flame which was grey would be *darker* than the average brightness of the surroundings. But if it is to *flame*, it must be considerably brighter than what lies around it. It cannot be both. So there cannot be a blazing grey or a dazzling grey. There are no grey filters for stage lighting. Just as a brown traffic light is impossible, so a grey flare would be invisible in darkness – it would not *flare*.

(v) Why is grey a neutral colour? The puzzle proposition that grey is neutral follows from the definition. For a deeper understanding of the assumptions needed to facilitate this deduction, Fig. IV can be compared with Figs. I and II (Chapter 2). The question must not, however, be taken in metaphysical fashion as the search for some detachable 'factor' which *makes* grey a neutral colour (as something makes or causes paper to yellow with age), in the absence of which grey would revert to the natural chromatic state. For grey is essentially achromatic. It could not be chromatic. The question should rather be what the basis of this essential truth in fact is, that is, in what the achromaticity of grey consists. For chromatically coloured objects, the significant fact is that the reflectance curves are not flat. Vermilion red cannot take on a colour whose curve rises steeply in the green, and vice versa. But the flat curve of Fig. IV can rise or fall slightly in the direction of any colour. Like black and white, grey does not selectively alter the incident light. The achromatic colours, one might say, are *neutral with respect to the differences between chromatic illuminant colours*, and they react in the same type of way to all of them. By contrast, the chromatic colours of bodies cause them to darken in lights of some colours but not in lights of others. This is why grey 'can take on the hue of any other colour' (vi), like white but not like black.

(vii) 'Grey is not poorly illuminated white . . .' What is important for our grey seeing as for our white seeing is not the absolute quantity of reflected light, but rather the loss (or lack of loss) of light caused by the object, the difference between the illumination to which we have adapted and the reflected light. This difference is selective or unselective darkening. Provided our eyes are adapted, we are perfectly well able to distinguish an area which reflects a small quantity of a powerful light from one which reflects a large quantity of a weak light.

In this way, dark green is not poorly illuminated light green, because the relationship of the two colours to the illumination is different, even though a painter might use the dark green to paint a poorly illuminated light green. Wittgenstein's analogy is exact.

(viii) Why is there no grey light? If an area is relatively dark or a shadow, it cannot also be an area of light or a light source. It does reflect some light – it is not black – but in quantity it must be relatively little and in quality it must be neutral. The illuminated black spot suspended in mid air appears grey because it seems to be reflecting a relatively high proportion of the illumination. And since grey is a sort of shadow, there cannot be a *beam* or a beaming of grey light. A beam is essentially a radiance, as eyes or a face are said to beam. A radiance here is a kind of shining.

What then are we to make of the familiar phrase 'the grey light of the morning', 'the cold grey light of the dawn', or Robert Peary's reference to 'the gray and shadowless light' in *The North Pole*? Here what is referred to, perhaps, is the general aspect, or in a loose sense the appearance of the *day*, rather than a photometric and colorimetric determination of the illuminant. For Peary writes a few sentences later, 'Notwithstanding the grayness of the day and the melancholy aspect of the surrounding world, by some strange shift of feeling the fear of the leads had fallen from me completely.' There is also an element of comparison involved. On an overcast morning, the dawn light will be cold and grey in the sense that it gives an overall impression of *shade* compared with a sunny midday. Of the *sky* (cf. the light, the day, the clouds, the weather, the last of which Peary describes as overcast) Peary writes that it 'was a colorless pall gradually deepening to almost black at the horizon.'

Earlier we noted that Irvin Rock was prepared to deny that illuminants are properly white or grey. They should, he claims, be described as emitting a particular quantity of achromatic light:

> The same is true about the sky when it is thoroughly overcast. Poetically it may be called gray, but to the author it does not look like a gray *surface*. Rather it looks luminous and has a particular brightness. Another example is a situation that provides uniform stimulation throughout the entire field, a Ganzfeld ... in which no contours whatever are visible. The observer has the impression of looking into a diffuse, three-dimensional fog ... and regardless of the intensity of light, the field never looks a shade of gray. Rather, it looks dim or bright. A completely dark room can be considered to be a special case of a Ganzfeld, and a dark room looks dark, not black. Darkness is the experience correlated with the absence of light, but this is not true of blackness. For the colour black to be experienced, certain specific conditions of contrasting luminances must obtain.[5]

Rock goes on to claim that where there are no surfaces of objects, there can be no white, grey or black, but only

[5] Rock, *An Introduction to Perception*, p. 503.

more or less brightness. Thus the light of a detuned television screen would not for him count as grey, and nor would a luminous sky. Perhaps also something is luminous only in a particular context, lacking in the case of a television screen in a darkened room. Here there is a harsh, flickering brilliance rather than a luminousness or a glow. In the case of the sky, there is also the fact that one could say that *to the extent that* it is luminous, it is not grey; so that the most nearly luminous sky would be the *lightest* grey. The light grey which is the grey the sky is when it could be called luminous is a grey which is very close to a white. Perhaps the nearest analogy is with the Eigengrau, or brain grey, which is perhaps better described as indeterminate in colour. This is not by itself sufficient to make it grey. It could, for example, be indeterminate as between black, grey and white.

So we can, if we wish, preserve Wittgenstein's claim that a grey light is impossible. One thing is clear: a light cast by a source which was itself a solid dark grey could not illuminate or *light* anything. 'A weak white light is not a grey light',[6] because a grey light would have to be not only weak but relatively dark. Accordingly, it could not be white.

(ix) The same principles are responsible for the impossibility of a planet appearing light grey. Wittgenstein does not, presumably, mean that a planet could not have a grey surface, or that a sphere of whatever size could not be painted a dull light grey. What he means is that a planet or star visible *as from the Earth in a night sky* could not or would not appear grey. For here the planet is a bright point light source in darkness. It cannot simultaneously be a dull area.

4.4 Concluding Remarks

Wittgenstein writes at *Remarks on Colour* I 49,

> Of two places in my surroundings which I can *see* in one sense as being the same colour, in another sense, the one can seem to me white and the other grey.

[6] *Remarks on Colour*, III 218.

> To me in one context this colour is white in a poor light, in another it is grey in a good light.
>
> These are propositions about the concepts 'white' and 'grey'.

This exploits a fascinating indeterminacy which is peculiarly related to grey; certainly it is related to the Eigengrau. Wittgenstein says in the next remark that it would be absurd to call the glazed shining white bucket in front of me 'grey', 'or to say, "I really see a light grey",' – even if grey were used to paint it, he might have added. In the context of the neutral luminous field of the Eigengrau after-images can be 'seen as' either light *sources* on a dark ground, or as coloured *darkenings* of a luminous illuminated field or in a luminous illuminated field which is bright rather than grey.

One thing does emerge clearly from the attempt to define grey. It is easy enough to say 'It is not the same thing to say: the impression of white or grey comes about under such-and-such conditions (causal) and: it is an impression in a certain context of colours and forms.'[7] It is another matter to sustain this sharp distinction in the face of the obvious fact that our concept of grey is what it is because of the anomalous behaviour, not of grey sensations or impressions, but of the colour grey. If the psychology of perception were restricted to references to sensations, perceptions, and impressions, then the distinction between logic and psychology drawn by Wittgenstein would have to be accepted. But why should we suppose that psychology must make do with these primitive and treacherous terms, or with the bad metaphysics which is suggested by the idea that they refer to 'entities' whose 'behaviour' is described using physical idioms of causation and interaction?

[7] *Remarks on Colour*, I 51.

5

Black

Know what whyte is, it is soon perceuved
what blacke is.

OED 1525

In Chapter 2 I showed how the puzzles which Wittgenstein raises about white in the *Remarks on Colour* can be answered by means of a definition of something white. Something is white if it does not significantly darken any light incident upon it, in the sense that the spectrum of light incident upon it is the same as the spectrum of light reflected from it, near enough. This is why white paper is used for water-colour painting and white screens for colour experiments. They 'take' respectively the colour of the paint and the colour of the light reflected from them. An equivalent and alternative formulation is that something is white if it reflects a high enough proportion of the light incident upon it, regardless of the colour of this light. In the present chapter I propose a definition, parallel to those given of the other colours, which solves the puzzles Wittgenstein raises about black.

Someone who knows what white is, in the sense of knowing what it is that white is, should have no difficulty guessing what, in the same sense, black is. Something is black, or, more strictly, is black coloured, if it more or less completely darkens any light incident upon it (Fig. V). Or something is black if it refuses to reflect light of any colour. It robs the incident light of its *entire* spectrum, as opposed to preserving the entire spectrum, as white does, or selectively preserving it, as objects and materials of other colours do.

This suggests that black will have certain properties in common with white. Black, like white, is in a certain sense the absence of a colour or of colour. Yet it can also be counted a colour in good standing. For 'If we had a checked wall-paper with red, blue, green, yellow, black and white squares we would not be inclined to say that it is made up of two kinds of parts, of "coloured" and, say, "uncoloured" ones.'[1]

Just as white cannot be blackish or dark (there are no dark whites, though some whites are darker than others), so black cannot be whitish or light. Nor can there be brilliant, bright or *pale* black. Though black can be the colour of something with a glossy surface, this is due to the fact that there are *two* surfaces present, or a combination of two distinct physical actions, one producing the black and the other on top of it producing the gloss effect.

One could say in parallel with *Remarks on Colour* II 6 ('Isn't white that which does away with darkness?') that black is that which does away with light. 'A "blackish" colour – e.g. grey – doesn't "shine".'[2] A shining is light surrounded by less light. A black light would be darker than what lies around it – this *follows* from the proposition that it is black – and brighter than what lies around it – it is a light. If something is brighter than what lies around it, it is not darker than what lies around it. So a black light would be both darker than what lies around it, and not darker than what lies around it. This is a contradiction.

Another way of showing the impossibility of black light involves the transformation of the curves of Fig. V into those of Fig. VI. Fig. VI shows how a vermilion light is one which darkens the blue, cyan and green samples, so that they turn black, and lights up yellow and red samples, which retain their colour, and appear more or less as they would in 'white' light. A white light lights up samples of all colours. A black light, however, would darken a sample of any colour. So it would not light up anything coloured. Accordingly it would not be a *light*.

Figs. V and VI also show why black takes away the *luminosity* of a colour,[3] and the brightness of colour.[4] The

[1] *Remarks on Colour*, III 37.
[2] Ibid., I 55.
[3] Ibid., III 156.
[4] Ibid.

Fig. V

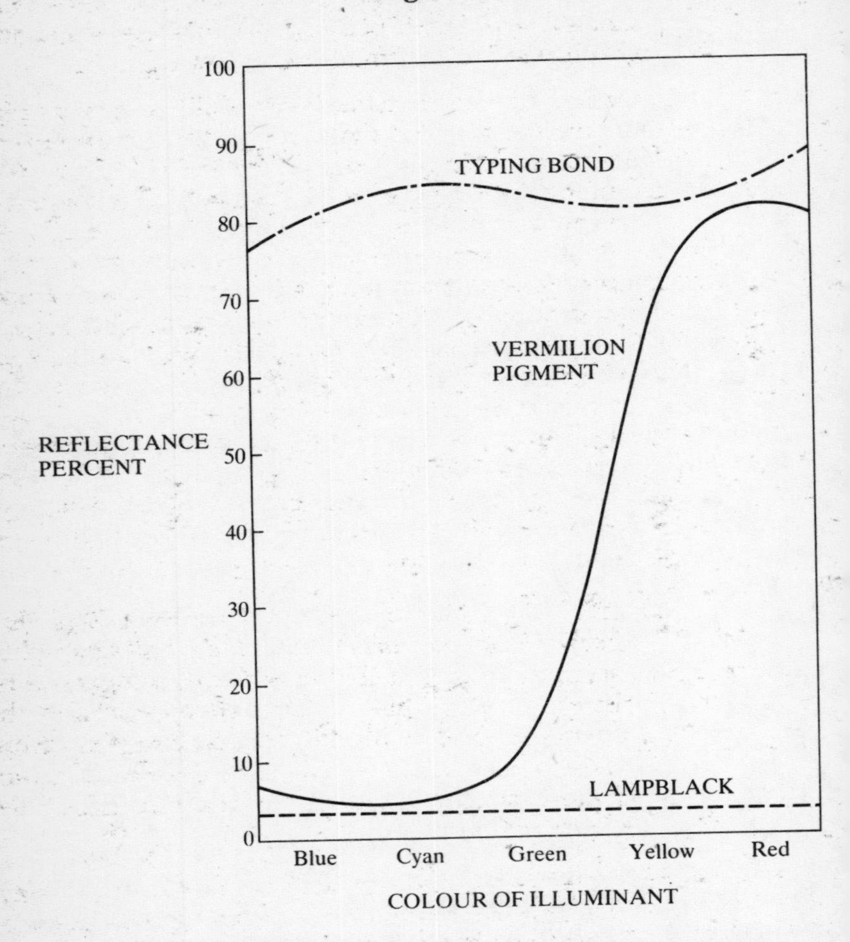

SPECTROPHOTOMETRIC CURVE FOR LAMPBLACK AND TWO OTHER SUBSTANCES

Fig. VI

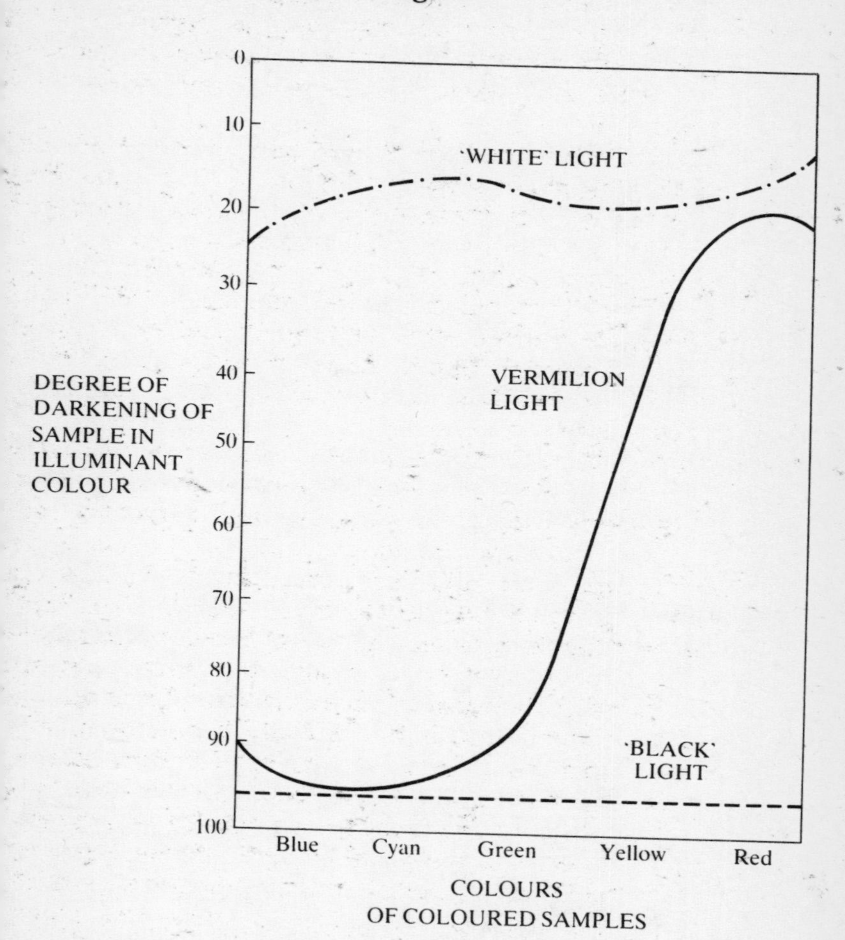

TRANSFORMATION OF SPECTROPHOTOMETRIC CURVE FOR LAMPBLACK AND TWO OTHER SUBSTANCES

first of these terms is strictly physical, the second phenomenal or psychological. But Wittgenstein is probably using 'luminosity' in a non-technical sense to mean *luminousness*, a sort of glow, a brightness, so to speak, in context. The spectrophotometric analysis given in the figures also gives a clue to the fact that, as Wittgenstein says, green is drowned in a black mirror,[5] but white is not, and to the fact that 'We say "deep black" but not "deep white".'[6] With black there must be three-dimensional depth. White things, on the other hand, must have defined surfaces to reflect or fail to reflect or darken the incident light in the sense given above. This anti-physicalist physical analysis also explains why black should be the darkest colour,[7] and therefore why it cannot be brilliant or bright.

Thus Wittgenstein is quite right to ask,

> Yes, suppose even that things only radiated their colours when, in our sense, *no* light fell upon them – when, for example, the sky were *black*? Couldn't we then say: only in black light do the full colours appear (to us)?[8]

He is also right to answer,

> But wouldn't there be a contradiction here?[9]

The contradiction is in the notion of the sky as a *black light*. With the account given above, we have two commensurate and 'contradictory' terms ('darker' and 'not darker' or 'brighter') in the analysis. Furthermore, it is wrong to say that under these or any other conditions things would radiate their colours, like phantasms. What they radiate is coloured *light*. It could of course be imagined that things appeared in their *true* colours only in the darkness, perhaps phosphorescently. But plainly what is being imagined here is not a black illumination or *light*.

This brings us to the important distinction between darkness and black, which arises in two distinct ways. (1)

[5] Ibid., III 238.
[6] Ibid., III 156.
[7] Ibid.
[8] Ibid., II 19.
[9] Ibid.

'"Dark" and "blackish" are not the same concept.'[10] What is the difference? 'It is not true that a darker colour is at the same time a more blackish one.'[11] An example would be yellow. A saturated yellow, Wittgenstein points out at *Remarks on Colour* III 70, is darker but not more blackish than a whitish yellow. (2) There is also a difference between blackness and darkness in three-dimensional contexts, in which what is dark or black is a physical object rather than a colour.

Thus,

> A ruby can appear dark red when one looks through it, but if it's clear it cannot be blackish red. The painter may depict it by means of a blackish red patch, but in the picture this patch will not have a blackish red effect. It is seen as having depth, just as the plane appears to be three-dimensional.[12]

This suggests that darkness could be regarded as blackness with the dimension of depth. But why should this be? The key difference between black and darkness lies in their relation to the illumination. Black is a surface colour, Wittgenstein declares at *Remarks on Colour* III 70 and III 156. Yet this is plainly not crucial. The real difference between blackness and darkness is that a black area *changes* an already present illumination, and it is surrounded by coloured things bathed in the same illumination. Darkness, on the other hand, is the *absence* of any illumination, or the lowering of an existing illumination, when it is blocked off or otherwise prevented from illuminating. Rooms in a house can be dark even though nothing in them is blackish or black; and certainly the illumination is not blackish, for the reasons given. What is important for black is the failure to reflect the illumination. What is important for darkness is the absence of illumination.

Under the influence of the wavelength theory, one can see how black must be regarded as the absence of colour, just as white is regarded as the sum of all colours. This too

[10] Ibid., III 104, cf. III 272.
[11] Ibid., III 70.
[12] Ibid., III 237.

simple picture overlooks the facts (1) that the superposition of *any* two complementary coloured lights produces white, though only on a white surface, (2) that the *colour* white cannot in any sense be regarded as the sum of all the other colours, although – a quite different matter – 'white' light results from the superposition of *lights* of complementary colours, and (3) that there is, as we saw, a perfectly good sense in which black is a colour in good standing. The mistakes are inevitable if we have the idea that colour is a *phenomenon* which has causes, and that these causes are stimuli conceived as light reflected from the coloured object. Then with *no* stimulation there is no colour phenomenon, no stimulus for black. So black is not a colour. But stimulation *is* required for black, and we should instead think of the colour in relation to the colour balance or spectrum of the illumination over the whole visible scene.

> Equip yourself with a lamp that is controlled by a dimmer. Go into a darkened room, slowly turn up the dimmer, and look at the contents. Notice that when you look at them under conditions of very dim light, the gray range is tightly compressed, with little visible lightness difference between the darkest and lightest objects. But as the light increases, the gray range expands in *both* directions: not only do the whites look whiter, the blacks look blacker. An increase in the total amount of *light* has increased blackness.[13]

For black, light or illumination is needed. For darkness, to which the dim greys are closer than the solid blacks in this respect, what is needed is a diminution or *lack* of illumination. Black is not *dim* in the same way that grey is.

One of Wittgenstein's purposes was to undermine the empiricist-inspired idea that the content of visual experience is a mosaic of colour-spots. He went further, to root out the very notion of a visual experience with content. He objected to the idea of content, of a *this*. The example of black suits this purpose. With black comes the relation to the illumination, and also the relation to darkness. What we are aware of when we are aware of black things or dark things must be understood, at the least, in the

[13] C.L. Hardin, 'Color and Illusion', unpublished MS, 1988, p. 6 (my italics).

three-dimensional context. This context is not the two-dimensional mosaic, and colour concepts for black and darkness cannot be derived from the mosaic, only in part because the mosaic has no relation to an illumination or only a fixed relation.

Such at any rate is *one* interpretation of the point of *Remarks on Colour* I 81 and certain passages in the later parts of *Remarks on Colour* III. 'Can one describe to a blind person what it's like to *see*?' Or I 86: 'Could a psychology textbook contain the sentence, "There are people who see?"? Would this be wrong?' What the blind man lacks is not *consciousness* as a kind of mosaic of spots of light. What am I believing in when I believe that a man is blind, to borrow from *Philosophical Investigations* 422?

> What am I believing in when I believe that men have souls? What am I believing in, when I believe that this substance contains two carbon rings? In both cases there is a picture in the foreground, but the sense lies far in the background; that is, the application of the picture is not easy to survey.
>
> One could also say of our *notion* of blackness, the colour black.
>
> The picture is *there*; and I do not dispute its *correctness*. But *what* is its application? Think of the picture of blindness as a darkness in the soul or in the head of the blind man.

6

Red and Green, and Physicalism

> An inadequate truth works on awhile; then, in place of full light, behold, a dazzling falsity steps in.
>
> GOETHE

> The fault of the analyses did not therefore appear to consist in saying too little – it did not seem that they failed by omitting to refer to some extra entity, for example – if anything, they seemed to provide too detailed an account of what was said in the analysandum.
>
> J.O. URMSON, *Philosophical Analysis: Its Development Between the Two World Wars*

6.1 Introduction

This chapter is concerned with the old question why nothing can be red (all over) and green (all over) at the same time. The answer is prepared by a discussion of the 'physicalist theory of colour'. The different question why there cannot be such a colour as reddish green is taken up in Chapter 7. There is a huge and rather baffling literature on the so-called predicate incompatibility problem. My purpose in this chapter is not to survey this literature. Instead, I shall assume that what has been said so far about white, brown and grey is correct, and offer a parallel solution for the red-green problem. It is worth trying to get a uniform account for all of Wittgenstein's puzzles. All by itself this would suggest something important about the nature of colours.

We saw earlier that the definition of whiteness could be given a phenomenalist interpretation. One advantage of this procedure is that it allows the solution to Wittgenstein's puzzles given in the case of whiteness to be extended to the puzzles about the other colours. According to Rock,

> In both cases the effect is a function of the relationship of the stimulus reaching the eye from one region to that reaching the eye from an adjacent or surrounding region, and in both cases it is unnecessary to refer to cognitive processes entailing the taking into account of illumination. However, the analogy stops there, because in the case of neutral colour, the hypothesis is that the ratio of intensities determines the perceived color, whereas in the case of chromatic color the notion of ratio of wavelengths does not seem appropriate.[1]

The general strategy I shall pursue in this chapter will be to look for an analogy which makes the principles governing the achromatic colours treated so far applicable to the chromatic colours. This will not involve a ratio of wavelengths, and it must be stressed that there is indeed no analogy here. The proposed ratios will involve not the actual wavelengths of the spectral composition of reflected light, but the potential colours of the illumination which would be darkened by the substance whose colour is being defined.

6.2 Preliminary Refutation of Physicalism

David Armstrong writes that, 'if we ask what in fact colours are, the physicalist reduction of these properties to light-emissions of different wavelengths' promises to produce the 'logical characteristics' required both for his theory of universals and for the solution to the predicate incompatibility problem.[2] Armstrong has also confusingly identified colours with properties which he calls, in a 'simplified assumption', 'grid grains', so that a red surface, for example, is 'a relatively fine-grained grid.'[3] A grid cannot be fine-grained and coarse-grained all over at the same time . . .' Armstrong

[1] *Introduction to Perception*, p. 549.
[2] D.M. Armstrong, *A Theory of Universals* II: *Universals and Scientific Realism* (Cambridge: Cambridge University Press, 1978), pp. 126 – 127.
[3] D.M. Armstrong, *A Materialist Theory of Mind* (New York: Routledge, 1968), p. 279.

does not say *why* it cannot; here we have the new philosophical problem of texture incompatibility. He claims that 'the history of philosophical discussion of this problem shows that a satisfactory answer is not easily found. It is a powerful argument for our view that it solves the problem.'[4]

I believe that I can show that the physicalist reduction cannot provide a solution to the incompatibility problem because it provides the wrong logical characteristics, and that the right logical characteristics are given by a theory of a quite different kind. At its most general, the complaint I have is that the logical complexity of a spectrum arranged on the basis of increases in frequency is too simple to match the actual structure of colour incompatibility – leaving aside related problems of the same kind as those discussed in Chapters 2–4.

Consider the yellow of the banana I shall eat for lunch. This yellow coloured object is reflecting light of a number of different wavelengths. If it reflected only the so-called 'yellow' wavelengths, about 3 or 4 per cent of the total, and absorbed all the others, including the 'red' and the 'green', it would be a *black* object (see Fig. I, p. 32). The yellow of the banana has to be reduced to a light emission of a number of wavelengths which are on Armstrong's theory actually incompatible, including amounts of the 'red' and 'green' which are actually *greater* than the yellow. The banana reflects 'red' and 'green' Rays more copiously than it does 'yellow' ones. At the *physical* level of wavelength of light emission, the banana is 'red' and 'green' and 'yellow' coloured. It is multi-coloured incompatible colours. If Rays are what colours are, then an object which is yellow by virtue of its reflecting *only* 'red' and 'green' Rays (those monochromatic Rays which taken singly appear red and green) will *be* red and green, and therefore incompatible with itself.

Consider also the position of white. White objects reflect light of any suitable pair of wavelengths or of many wavelengths. On the physicalist theory, the complex spec-

[4] Ibid., p. 280.

tral composition of 'white' light will make the white objects incompatible with themselves. A cyan at 494 nm. is incompatible with the complementary red at 700 nm. Yet the two together make white. How can it be that it is exactly the incompatibility of these two wavelength values which *explains* the incompatibility of red and green, but does *not* prevent a white surface from being uniformly one colour? How is white composed of the incompatible 494 nm. and 700 nm. emissions possible at all?

There is no such thing as monochromatic white, so any white sample will reflect light of at least one pair of incompatible wavelength values. It will be impossible to *state* the compatibility of white composed of the 494 nm. and 700 nm. with an identical white composed of complementary yellow at 570 nm. and blue at 436 nm. The same problems arise for the nonspectral purples, browns and other 'physically mixed' colours. The appearance of brown, as we saw, depends not on wavelength but on whether it appears as a light or a shadow, and the same is true of grey. These colours are physically incompatible with themselves on Armstrong's theory. Or if brown is somehow not incompatible with itself, how can a wavelength incompatibility of precisely this type explain the impossibility of something simultaneously red and green? These examples suggest that it would generally be better to distinguish colour and spectral composition, as has been done in colour science since Newton's celebrated distinction between the Rays and the sensations they stir up in the sensorium.

I agree with Armstrong that there is a simple connection between 'what colours are' and the solution to the incompatibility problem. My view is that incompatible things are incompatible because of what they are, and that the incompatibility problem is a test which any theory of 'what colours are' must pass. Armstrong's physicalist reduction fails it. So colours must be something other than light emissions of different wavelengths.

These points are brought against the physicalist solution to the problem of colour incompatibility. I propose now to outline a number of objections to the programme of reduction as a whole on which this solution is based. These objections all make the same kind of point. They belong with the other empirical objections to identifying secondary qualities with physical properties given in Chapter 3 of *A Materialist Theory of Mind*.[5] I shall try to show that there are important and true statements in the theory to be reduced (statements about colour and colours, hue, saturation, brightness, brilliance, etc., and statements using these terms) which become false when these psychological predicates are replaced by the physical predicates of the reducing theory.

6.3 The Full Refutation of Physicalism

It is a necessary condition for reduction in Tarski and Woodger's sense[6] that statements which are true in the reduced theory should be preserved in the reducing theory. Armstrong does not say what he understands by reduction, but I shall assume that if I can show that this condition is not satisfied, the reduction must fail. The absence of biconditionals relating each primitive predicate of the reduced theory to some, possibly complex predicate of the reducing theory means that the physical theory does not even *interpret* the psychological theory in Tarski and Woodger's sense. Without fixing the sense of 'reduces', it is hardly worth discussing the reducibility of colour theory to physics, or to anything else.[7]

[5] Ibid.

[6] J.H. Woodger, *Biology and Language* (Cambridge: Cambridge University Press, 1952), pp. 271–272.

[7] Kenneth Schafner outlines some alternative senses of reduction in his well-known article 'Approaches to Reduction', in *Philosophy of Science* **34** (1967), pp 137–147. Some of these make very little sense in the colour case. In others the reduction simply fails. In the Kemeny–Oppenheim paradigm, it is false (condition 2) that the observational data associated with the psychological part of colour theory can be *explained* by the physical theory. In the weaker version of Popper–Feyerabend–Kuhn, the psychological theory cannot be explained by the physical theory even in the minimal sense that the physical theory yields predictions which are 'very close' to the predictions of the psychological theory. Indeed, the physical theory makes *no* predictions without the psychological theory. As for the Suppes criterion, it is just a brute fact that there is no isomorphic model of colour theory in the physics of light. Psychological colour space is three-dimensional (hue, saturation and brightness) and it cannot be derived from the two-dimensional psychophysical chromaticity diagram, let alone the stimulus itself. There is also

We know that a colour (strictly, perhaps, an absolutely determinate hue or colloquially a shade of colour) can be saturated or unsaturated. But it is some kind of nonsense – a category mistake – to say that a light emission of such and such wavelength is saturated or unsaturated. The physicalist must take steps to replace this absolutely essential pair of predicates in the reducing theory. The reduction of colours requires not only a reduction of the colour words 'red', 'green' etc., but the additional reduction of the terms 'saturation' to 'colorimetric purity', and 'brightness' to 'luminance'. The colorimetric purity of a sample is the ratio of the amount of the spectrum component to the sum of the spectrum component and the daylight component. The dominant wavelength of a colour is the wavelength of that part of the spectrum required to be mixed with some chosen light (daylight) to produce the colour, so that in our earlier example a colour may contain no 'yellow' Rays, but only 'red' and 'green' ones, and still have a dominant wavelength of 580 nm. because it matches monochromatic yellow of that wavelength. 'Dominant wavelength' is a psychophysical term, not a physical term, and the same is true of 'luminance' and 'colorimetric purity'. This is evident from the fact that the dominant wavelengths of the nonspectral purples, which cannot be monochromatic, are defined as the (physical) complementaries of colours with dominant wavelengths in the green. And generally the term relates a physical magnitude to what is seen by the observer. It links colours of many different wavelengths to the wavelengths of the monochromatic standard.

In what follows I consider the replacement of the psychological predicates by psychophysical predicates. The further reduction of the psychophysical predicates to physical predicates ('spectral selectivity' and 'spectral radiance') represents an additional problem which I shall not consider. I shall concentrate on the first psychophysical hurdle. If the reduction fails here, it is irrelevant whether another reduction could be performed on the psychophysical terms.

---

doubt as to whether even the addition of physiological functions could effect the transformation.

The total project of reducing colours to light emissions of specified wavelength would already have failed. And it does fail. Even at the first stage it founders on some well-known empirical facts. I shall also offer a mixed bag of examples which show the inadequacy of the physicalist reduction in the wider setting of colour perception as a whole.

(i) Hue and dominant wavelength: We know that there is a variation of hue with saturation. The addition of white to a blue, for example, makes the colour not only whiter but redder. The hue shifts from blue into mauve. The dominant wavelength does not change, so that mauves which we see may correspond to light emissions whose dominant wavelength lies in the blue. Substituting 'dominant wavelength' for 'hue' and 'colorimetric purity' for 'saturation', we get the statement that there is a variation of dominant wavelength with colorimetric purity. This statement is either false or nonsensical. The difficulty is perfectly general. The lines of constant hue in a chromaticity diagram diverge from the lines of constant wavelength throughout, not just in the blue.

(ii) Adaptation and colour constancy: The colour of the light reflected into the eye from some chosen part of the visible scene will vary enormously through the day, as colour photographs will show. But the objects in the visible scene do not change colour, nor do we see a change. This is due to adaptation to the colour of the illumination. The reducing theory records a change in the colours of objects which the reduced theory denies.

(iii) Hue and Brightness: Consider the variation of brightness with hue. Fig. VII shows brightness against colour for an equal energy spectrum, with the yellows and greens showing the greatest brightness. Colours of the same physical energy differ in brightness. Brightness matches for different colours can only be achieved with different energies. But in the physical theory the colours in Fig. VII do not vary in energy.

Fig. VII

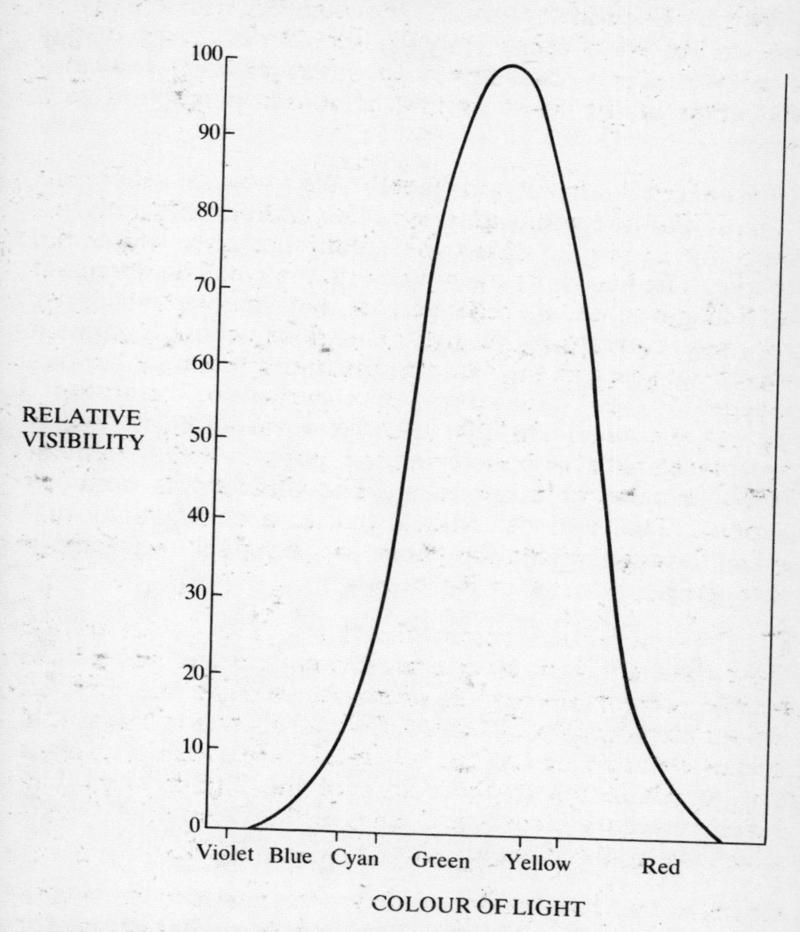

RELATIVE VISIBILITY (BRIGHTNESS) FOR DAYLIGHT VISION

(iv) Unitary Hues: In spite of the fact that the hue circuit can be arranged so that the spacing between the hues is equal (each adjacent hue is just noticeably different from its neighbours), at certain points there seems to be change of another kind. These points are the so-called non-unitary hues. The hues that lie between the unitary hues can be described in terms of them, so that a turquoise, for example, can be described as a blue-green. But the unitary hues cannot be described in terms of the unitary hues. Red cannot be called a purple-orange. To what predicate in the reducing theory will 'unitary' reduce? Consider the statement that unitary green is neither yellowish nor bluish. Should the reducing theory state that a monochromatic emission of 497 nm. (unitary green) does not contain a 'trace', as the psychologists say, of 570 nm. (unitary yellow) or of 470 nm. (unitary blue)? But this is trivially true, since the original emission was monochromatic. If this is what 'unitary' means in the physical theory, then all monochromatic colours are unitary. Furthermore, unitary red is nonspectral. It cannot be produced by light of a single wavelength, but only by a mixture. So there is no physical sense in which the unitary hues are purer than the others. I have some sympathy with the doctrine of *Remarks on Colour* that the status of the unitary hues is a fundamentally logical one. On the other hand there is general agreement in colour science that the explanation of the positions of the unitary hues is to be found in post-retinal coding. What is certain is that it cannot be found in the physics of the stimulus. Perhaps there is no explanation why the unitary hues fall where they do. There is no explanation of the pattern in which prime numbers fall on the number line, and there is no formula which generates only primes. They are 'mathematical danglers'. In any event, we have a clear truth in the reduced theory ('Such-and-such a hue is unitary') which is not preserved in the reducing theory, and cannot even be stated in it.

(v) Primaries: Exactly the same problem arises with the primaries of additive or light mixing and the primaries of subtractive or pigment mixing. Take the three primaries of additive mixing, red, green, and violet. How will the term 'additive primary' stand in the physical theory? The additive primaries have no physical distinction – at least in the physics of electromagnetic radiation. It will not do to say merely *which* colours are primary. What we need from the physical theory is a statement of *what* a primary is. To adapt a remark from Wittgenstein: a primary colour is not simply: *this*, or *this*, or *this*. We recognize or determine it in a different way.

(vi) Rate of hue change in the spectrum: In the course of a discussion of the fact that there is no 'simple or constant connection between the wave-lengths of pairs of complementary colours' (NB what will 'complementary' reduce to in the physical theory?), Helmholtz observes: 'At the ends of the spectrum the hue changes exceedingly slowly as compared with wavelength, whereas it changes very fast in the middle of the spectrum.'[8] What will be the corresponding statement in the reducing theory? It must be that dominant wavelength and ultimately wavelength of light emission change extremely slowly at the ends of the spectrum *as compared with wavelength*, whereas they change very fast in the middle.

(vii) Metamerism: Two samples of colour C^1 and C^2 are said to form a metameric pair when they match, but the light reflected from the coloured material differs in spectral composition. Isomers are samples which match but produce light of the same spectral composition. Let the light from C^1 be λ_1 and the light from C^2 be λ_2 in wavelength. Now there is a truth about C^1 which is not a truth about C^2. C^1 gives light of λ_1, C^2 does not. The response of the physicalist is to redefine the colour as a disjunctive class of light-emissions (call it C*), into which λ_1 and λ_2 fall. C* is the colour (λ_1 or λ_2 or λ_3 or λ_4 . . .).[9] If a colour is C* it can be either C^1 or C^2, and identity is preserved.

[8] Helmholtz, *Physiological Optics* II, pp. 126–127.

[9] J.J.C. Smart, 'Reports of Immediate Experience', *Synthese*, 22 (1970–1971), p. 356 ff., and also the same point made in his 'On Some Criticisms of a Physicalist Theory of Colours', C.Y. Cheng (ed.), *Philosophical Aspects of the Mind-Body Problem* (Honolulu: University of Hawaii Press, 1975).

But there is a difficulty. The move to C* destroys the distinction between metamers and isomers. C^1 and C^2 must not be in any way distinguishable, or they fall foul of Leibniz's Law and are not the same colour. But if they have the same (disjunctive) composition given in C*, they are not metamers – *they are isomers*. If on the other hand C^1 and C^2 are distinguished by spectral composition, they are not identical or not the same colour. So they are not metamers. Either way they are not metamers. What is needed here is the very thing the physicalist is dedicated to not giving – the *distinction* between colour and spectral composition, which is, after all, the basis of the concept of metamerism. This distinction should be combined with a recognition of the fact that a *colour* cannot be said to have a spectral composition. What has a spectral composition is the light reflected from a coloured material. But this is what the physicalist thinks a colour is.

(viii) Finally, Grassman's Law: Grassman's Third Law, which is the basis of modern colorimetry, states that 'Two colours, both of which have the same hue and the same proportion of intermixed white, also give identical mixed colours, no matter of what homogeneous colours they are composed.'[10] Of this law Judd writes, 'It means that we can deal with lights on the basis of their colours alone and without regard to spectral composition.'[11] How will these two statements stand in the reducing theory?

6.4 Further Criticism of Physicalism and a Reply

The relationship between colorimetry and the physics of radiation illustrates perfectly the celebrated commonsense conception of autonomous explanation advanced by Putnam and referred to in connection with white in Chapter 2. As we saw, Putnam thinks that particle mechanics and quantum electrodynamics will not explain why his square peg will not go into a round hole.

[10] H.G. Grassman, 'Theory of Compound Colors', in David L. MacAdam (ed.), *Sources in Color Science*, p. 60.
[11] D.B. Judd, *Color in Business, Science and Industry* (New York: Wiley, 1952), p. 48.

. . . if you are not 'hipped' on the idea that *the* explanation must be at the level of the ultimate constituents, and that in fact the explanation might have the property that *the ultimate constituents don't matter*, that *only the higher level structure matters*, then there is a very simple explanation here. The explanation is that the board is rigid, the peg is rigid, and as a matter of geometrical fact, the round hole is smaller than the peg, the square hole is bigger than the cross-section of the peg. The peg passes through the hole that is large enough to take its cross-section, and does not pass through the hole that is too small to take its cross-section. That is a correct explanation whether the peg consists of molecules, or continuous rigid substance, or whatever. (If one wanted to amplify the explanation, one might point out the geometrical fact that a square one inch high is bigger than a circle one inch across.)

Now, one can say that in this explanation certain *relevant structural features of the situation* are brought out. The geometrical features are brought out. It is *relevant* that a square one inch high is bigger than a circle one inch across. And the relationship between the size and shape of the peg and the size and shape of the holes is *relevant*. It is *relevant* that both the board and the peg are rigid under transportation. And nothing else is relevant. The same explanation will go in any world (whatever the microstructure) in which those *higher level* structural features are present. In that sense *this explanation is autonomous*. . . .

The explanation at the higher level brings out the relevant geometrical relationships. The lower level explanation *conceals* those laws. [my italics]

The fact is that we are much more interested in generalizing to other structures which are rigid and have various geometrical relations, than we are in generalizing to *the next peg that has exactly this molecular structure*. For the very good reason that there is not going to *be* a next peg that has exactly this molecular structure. So in terms of real life disciplines, real life ways of slicing up scientific problems, the higher level explanation is far more general, which is why it is *explanatory*.[12]

[12] Hilary Putnam, 'Philosophy and Our Mental Life' *Mind, Language and Reality*, 2, pp. 295–296.

In colorimetry a physical specification in terms of wavelength is irrelevant where a higher-level specification brings out the relevant relationships – if only because we need to 'generalize' to mixtures of colours with the same physiologically based chromaticity coordinates irrespective of spectral composition. There are not going to *be* two lights of *exactly* the same spectral composition.

Armstrong does not consider objections of exactly this type. But he does offer a reply to a somewhat similar objection. 'In recent years Smart has emphasized again and again that a huge and idiosyncratic variety of combinations of wave-length may all present exactly the same colour to the observer'[13] rather making it sound as if Smart had invented colorimetry. (A minor but irritating inexactness – do wave *lengths* present colour?) 'The simple correlations between colour and emitted wavelengths [sic] that philosophers hopefully assume to exist cannot be found.'[14] No philosopher acquainted with the history of colour science should ever have supposed such a thing. It is not some sort of curious *anomaly* that lights of a variety of wavelength can have the same colour. There is unity in this variety, but unfortunately for the physicalist it is a *physiological* unity.

Or rather, and this is a very important qualification, *if* the conceptual resources of our theory of colour are only wavelength of light and response, *then* the unity can only be physiological. Armstrong *accepts* the outline of such a theory, and denies the implication of colour in the response. Certainly in this realist instinct there is something correct. But what is really needed is a higher level and simpler *description of the physical*, of the kind given in Gibson's ecological optics. (I shall offer the outlines of such a description for red and green in a moment.)

Armstrong's reply to the Smart objection is this:

> One possibility here is that all these different combinations of wavelength may be instances that fall under some general formula. Such a formula would have to be one that did not

[13] *A Materialist Theory of Mind*, p. 288.
[14] Ibid.

> achieve its generality by the use of disjunctions to weld together artificially the diverse cases falling under the formula. Provided that such faking were avoided the formula could be as complicated as we please.[15]

And, says Armstrong, such a formula could be 'tested'! But he is clearly clutching at straws. There is not a shred of evidence for any such complicated formula, even though,

> Because of its complex and idiosyncratic nature, the physical property involved was not of any great importance to physics. But it would still be a perfectly real, even if ontologically insignificant, physical property. Now I know of no physical considerations that rule out such a possibility.[16]

This is a terribly weak conclusion: there are no known physical considerations which rule out the as yet undescribed formula. Without it, however, Armstrong says that 'we would have to conclude that colour is a pseudo-property: a sorting and classifying of surfaces by means of the eye that has no proper basis in physical reality' – all because it isn't physics! And where does this leave the solution to the problem of incompatibility which was such a powerful argument for Armstrong's view? The solution depends on the possible (!) existence of 'some general formula' about which we are told *nothing*, except that as far as Armstrong knows there are no considerations to rule it out. So there might be, or there might not be, something here which explains, or might explain, colour incompatibility. But this is something which can't sensibly be discussed until the formula is before us. Armstrong's 'powerful argument' has dissolved. And if it should turn out that there is no formula, then how is colour incompatibility explained? How can it be explained if colour is a 'pseudo-property'? How would one set about solving the new but equally pressing problem of pseudo-predicate incompatibility?

It should by now be evident how important it would be

[15] Ibid. Here there is a disagreement between Armstrong's physicalism, which does not allow disjunctive properties, and Smart's, which does. See note 9 of this chapter.

[16] p. 289.

to get from Armstrong a precise formulation of the relevant biconditionals, an axiomatization of both the reduced and reducing theories, or at least a clear account of what he takes them to state, and further guidance on the 'general formula'. The vague 'colours are light-emissions of different wavelengths' will just not do. It is this vagueness at crucial points which makes physicalism so unsatisfactory. I find the explanation of this in the fact that in the physicalist conception scientific theory is regarded as true in a uniform and unproblematic way, as if there were no deep philosophical and scientific problems *within* what colour science says, and not just in the application of the whole body of theory to the physical world. No justice is done to the fact that colour science grows out of common sense and experience at a number of points, and that the theoretical terms of physics which it uses cannot be simply substituted for the terms of observation without damage to the delicate relationship between fact and theory, and between physics and physiology, *within* colour science.

Armstrong is working with a vastly oversimple dichotomy. *Either* a colour is what he says it is, *or* it is a pseudo-property, illusion or mere appearance. But this is a dichotomy which is close to the heart of physicalism. Either propositions are propositions of physics, or they are pseudo-propositions. The failure of the biconditionals in (i)–(viii) has a simple cause. The whole purpose of the reduction is to cut off colour science at the stimulus. This makes unnecessary almost all of the important theoretical advances since Thomas Young. And on top of this, the stimulus itself is wrongly described. The fact is that within the sciences concerned with perception the stimulus is constantly being redescribed in new and more interesting ways, e.g. in connection with the so-called 'cues' for adaptation, texture gradients for depth perception, edges for colour vision, etc. It does not follow from these advances that the secondary qualities are not physical, or that they must be mental or in the head. There are other ways of being physical than possessing an essence of ultimate particles, and the mental as usually represented is just a slop of everything we don't understand. The remedy is not a denial of the mental, but a better understanding of it – and a better understanding of the physical. I have no confidence that physicalism can give either, since it apparently has nothing to bring to the delicate metaphysical problems except a synopsis of elementary textbook physics.

6.5 A General Formula for Physical Colours

My own view is that there is a 'general formula' of a certain kind under which the same colours, correlated with different frequencies, fall; but it is a formula concerned with light or illumination as this is perceived, and the darkening of the light by objects, which produces edges. It is not a complicated formula, and it has only a complicated and peripheral relationship to the spectral composition of the light. On this view, the theory of colour will not reduce to physics, not only because the idiosyncratic nature of colour prevents it, but because the received physical account is *wrong*.

In a conception put forward by Wilson and Brocklebank,[17] the significant property for colour perception of e.g. yellow objects is not the colour or spectral composition of the light they reflect, but rather the colour of the light they *don't* reflect, namely light whose colour is *blue*. More precisely, it is the small amount of blue light *relative to light of the other colours* which they reflect which determines their colour. What counts is the colour of the light they darken or absorb. It is a significant fact that the dyer, who is concerned not with the colours of lights (as Newton was), but with the colours of objects and materials, typically works with the complementary or *absorption* spectrum as his tool, not with the reflection spectrum. The yellow of the objects is seen as the way in which the light is *darkened* away from the blue. This was also apparently Goethe's basic conception. An illustration from Wilson and Brocklebank may help. Let a yellow filter be held up close

[17] M.H. Wilson and R.W. Brocklebank, 'The Phenomenon of Coloured Shadows and its Significance for Colour Perception', *Die Farbe* **2** 1/6 (1962), pp. 143–146, 'Colour Is Where You See It', *Internationale Farbtagung* (Lucerne: Tagungsbericht, 1965), pp. 991–1001, 'Goethe's Colour Experiments', pp. 3–12.

to a white screen onto which blue light is projected. The light falling onto the white screen from the filter gives a *shadow* which is definitely *yellow*. Let the yellow filter be brought back until it is right in front of the blue filter on the projector. Now the whole screen is *green*. Why? When we see the image or shadow of the yellow filter on the screen, we are aware that the filter darkens the colour of the illumination, for we have adapted to this blue light and we can assess *departures* from it. This does not happen when the yellow is a component of the illumination itself. Here there is no change in the green illumination. In the first case the yellow is a darkening or shadow, a changing of the illumination, whereas in the second case it is a part of the unchanged light. In both cases the light reaching the eye is the same, but its relationship to that light is different.[18] It is exactly this type of phenomenon which underlies the definitions already given in Chapters 2–4.

A green object in this scheme is an object which *refuses* to reflect a significant proportion of red light relative to lights of the other colours, including green. It would be nice to be able to say that a green object absorbs a large proportion of the red light or red component of white light relative to the components of other colours. Then we would be able to say that an object cannot be simultaneously green (all over) and red (all over) because it cannot simultaneously absorb and not absorb a high proportion of the red component, and of the green component. But we cannot say this, because there is no *blue* component in *white* light.

We could say instead that it is the relative longwave lack ('red' lack) in the total energy balance which is the significant property. I also want to reject this, however, on the ground that it will not enable us to transfer whatever necessity is obtained via a psychophysical correlation to actual *colour* incompatibility.

So I want to take genuinely red light, and consider the potential failure of a class of objects to reflect a significant

[18] This point concerns the correct understanding of black. See W.D. Wright, 'The Perception of Blackness', *Die Farbe* **28** 3/6 (1980), pp. 161–166.

proportion of this light, relative to light of the other colours, as the significant mark of green objects.

Fig. VIII represents the facts of colour mixing with filters of different primary colours. The transmitted portion of the spectrum is shown in white, and the absorbed portion in black. (The phenomenalist interpretation of 'transmitted portion of the spectrum' would be 'colour of the light which is not darkened' – and so remains *light* and passes through the filter onto a screen or something of the sort.) The additive primaries are shown on the far left, the subtractive primaries on the right.[19] The coloured filters are ideal in the sense that the blue filter, for example, transmits *only* blue light. Hence the transmittance curve is vertical. This representation could be regarded as providing a 'logic' for colour mixing. In additive mixing, we could say that the rule is that light prevails or is dominant, so that light + dark = light, or, in the diagram, white + black = white. Thus from red and green we *add* the lights or white parts of the diagram, and leave over in darkness or black only the blue part of the spectrum, so the result is negative blue or yellow. Similarly for the subtractive mixing, dark prevails or is dominant, so that the rule is that light + dark = dark, or, in the diagrams, white + black = black. Thus from yellow (−blue) and cyan blue (−red) we get green. Note that the principle of dominance gives us what could only count as a *white* or zero spectrum from the additive mixing of blue and yellow. These principles demonstrate, as W.D. Wright observes, the 'complete harmony between the principles of additive and subtractive mixing'[20] or between lightening and darkening processes. From this point of view, which it must be remembered concerns only physical colours, a colour could be regarded as a neutral or white spectrum (white in the diagrams) more or less darkened in one part or another. Magenta, for example, is a spectrum from which green has been removed. This definition of a colour as the alteration of a complete spectrum is, I

[19] W.D. Wright, 'Colour and Its Measurement', *CIBA Review*, Manchester (1961–2), p. 13. I have phenomenalized Wright's transmission spectra diagrams by replacing the wavelength scale with a spectrum scale from blue to red.
[20] Ibid., p. 12.

Fig. VIII

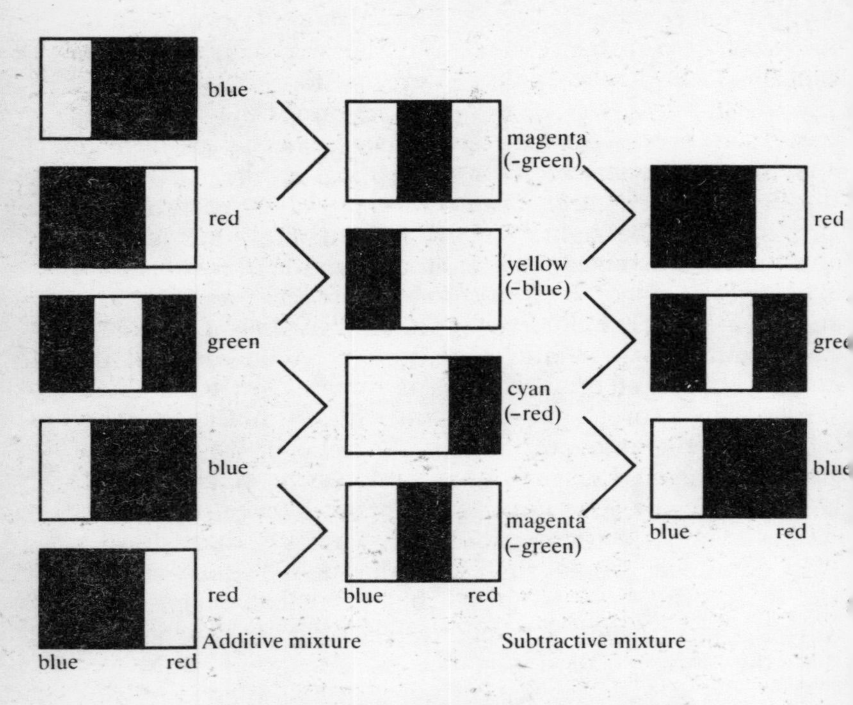

PRINCIPLES OF LIGHT AND PIGMENT MIXING

believe, significant from the point of view of the role of adaptation in colour perception. It is of course incomplete with respect to nonspectral colours, and in connection with the relation of colours to their viewing conditions. It avoids the need to refer to the *red* component in white light, which is nonexistent, or the 'red' component (longwave energy) in 'white' light, which still needs to be correlated with the colour red. We can interpret the 'red' potential of 'white' light as the disposition of colourless light to render red things red, other things being equal.

Now we can say that, provided it does not change colour, an object cannot both absorb or darken the relevant proportion of the red light relative to lights of the other colours, and absorb or darken the relevant proportion of green light relative to lights of the other colours, including red. A green object is an object which will absorb or darken almost all of any incident red light and reflect or not darken higher proportions of light of the other colours. The corresponding definition of a red object gives us a straightforward logical contradiction.[21] (It is a significant feature in this conception that with the afterimage of the green object the

[21] R.B. Braithwaite has told me of a conversation he had had with Friedrich Waismann about the status of the proposition 'This is red'. Waismann's view was that 'red' means 'red and not yellow and not all the other colours'. This is good Hegelianism. In the *Science of Logic*, Hegel writes that 'Something is in this relation to Other from its own nature and because Otherness is posited in it as its own movement; its Being-in-self comprehends negation, through which alone it now has affirmative existence ... it is just this cancellation of its Other that is something.' Hegel must surely have been thinking of colour cancellation when he wrote this passage. The trouble with Waismann's unmetaphysical version is that 'red' just doesn't *mean* 'red and not green and not all the other colours'. It is not part of the meaning of 'teapot' that it is not a coffee pot or a watering can or anything else. My solution to the incompatibility problem has the same form as Waismann's, but it differs in that it gives a real definition of the colours, not the meaning of the colour words. The essences exist so to speak only in relation to one another in the colour continuum, as Wittgenstein became aware after 1929 and as Hegel apparently already knew. But this is a real as well as a conceptual matter. It is significant as well as amusing that in practical colour contexts red is defined, as in Wright's ideal transmission spectra, as negative green. Thus in some sense of 'means', 'This blue is too green' means 'This negative yellow has a red lack.' Waismann's Hegelian holism is markedly inferior to the middle Wittgenstein's original. Against *Tractatus Logico-Philosophicus* 2.1512 and 2.15121, Wittgenstein said, 'If I say, for example, that this or that point in the visual field is *blue*, then I know not merely that, but also that this point is not green, nor red,

eye is positively calling out that there is a red lack in the object before it, and that it is in effect through making up the lack that we are informed of what colour the object is.)

This method of defining the colours of objects presupposes the ascription of colours to lights. The colour of an object is defined by reference to how that object changes the colour of the light. It might be thought that this procedure takes as primitive the very terms it is intended to define. The method can however be extended to the colours of lights. Clearly we cannot say that the redness of a light is its disposition to darken or fail to reflect green light. But the method used for defining object colour can be put in reverse, so to speak. We can say that a light is red if it is disposed to darken green objects, and that it is green if it is disposed to darken red objects. For greater accuracy, we must take the whole absorption spectrum into account, and read off the colour from the absorption profile. In Fig. I, for example, the typical red light will darken the grey card and the lampblack, but not the vermilion or the white, and it will also darken any green objects.

F.P. Ramsey wrote that there will always be circularity at the foundations of knowledge. This could be taken to mean that there is an *inevitable defect* in our knowledge. Wittgenstein perhaps took it this way, and then struggled to show that the inevitable defect was not after all a defect, by means of *philosophy*. In geometry, however, the inevitable defect is not a defect, but not for reasons given by philosophical argument. It is an expression of the most profound principle of projective geometry – the principle of duality. The principle states that every proposition about the elements in a projective plane stays true when 'point' and 'line' and various other pairs of words such as 'vertex' and 'side' are systematically exchanged.[22] Two such

---

nor yellow, etc. I have laid the entire colour scale against it at one go' (Friedrich Waismann, *Ludwig Wittgenstein and the Vienna Circle* (Oxford: Blackwell, 1979), p. 64).

[22] In *Projective Geometry* (Toronto, 1974), p. 25, H.M.S. Coxeter says that 'One of the most attractive features of projective geometry is the symmetry and economy with which it is endowed by the principle of duality: fifty detailed proofs may suffice to establish as many as a hundred theorems.'

propositions are called *duals* of one another, and can be written in parallel columns. Thus

Every line is incident with at least three points.	Every point is incident with at least three lines.

Or

If four points in a plane are joined in pairs by six distinct lines, they are called the vertices of a complete quadrangle, and the lines are its six sides.	If four lines in a plane meet by pairs in six distinct points, they are called the six sides of a complete quadrilateral, and the points are its six vertices.

Thus the quadrangle and the quadrilateral are dual configurations, and the dual of a triangle is a triangle; it is self dual.

In colour theory, complementary terms can be regarded as duals.

A green object refuses to reflect red light.	A red object refuses to reflect green light.

A medium grey could perhaps be regarded as self dual. But there are also what might be called *dual configurations*.

A green *object* darkens (refuses to reflect) red *light*.	A green *light* darkens red *objects*.

The *object* here is the dual of the illuminating light, and this duality produces the distinction between the two main types of colour mixing, of object colours or bodies and of lights.

The colours of bodies are governed by the rule of subtractive mixing. So we could represent the contradiction in the idea of an object simultaneously red and green as follows, with the dark portions of the spectrum representing the fact that the coloured object refuses to reflect light of that colour, and the light portions showing that the object does reflect light of that colour. We have:

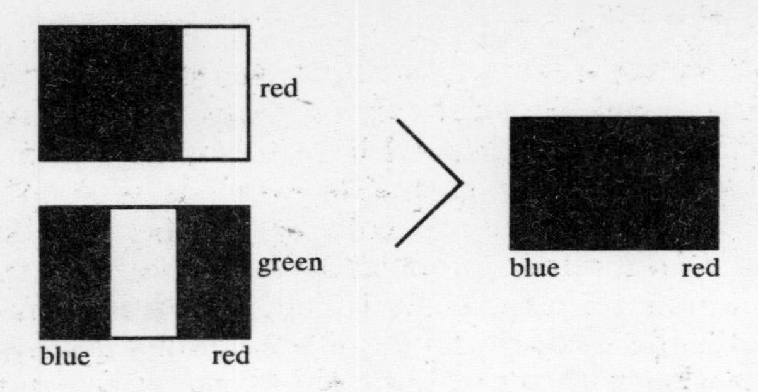

Note also that the same result would be obtained with magenta instead of red.

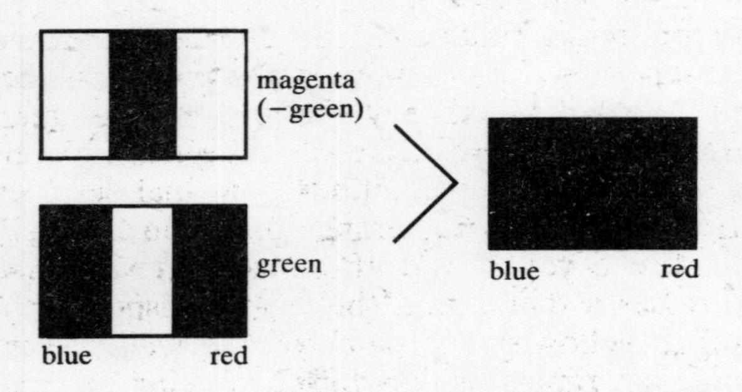

This shows that the relevant mark of red objects is their refusal to reflect green light. The diagrams show how what is offered, so to speak, in the way of colour by one spectrum (or colour) is withdrawn by the other, so that the result is the vanishing of colour, just as in a contradiction between two propositions which negate one another the result is a vanishing of information. The diagrams illustrate this for the two contradictory propositions in question. Note the similar relation in Fig. VI between yellow and blue.

6.6 Afterimages

I must now say a few words about why I do not think that the Wilson and Brocklebank property would serve in a physicalist reduction. What I have to say here also applies to white, brown and grey. The analysis is supposed to provide the essence of the colour, and the successful derivation of the puzzle proposition is an indicator that this has been done. But the analysis does not explain, nor could it explain, the incompatibility of the colours in physiological effects such as the coloured shadows effect and afterimage colour.

I think it is absolutely certain that afterimages do have colour (though it does not strike my ear as correct to say that they are *coloured*), and I think that only philosophers who have painted themselves into a corner in the philosophy of mind would ever have wished to deny such a flat empirical fact, however it should ultimately be analysed. The attempt to get rid of afterimages and afterimage colour has been made by J.J.C. Smart in *Philosophy and Scientific Realism*.[23] When I have a yellow orange afterimage, 'What is going on in me is like what is going on in me when my eyes are open, the lighting is normal, etc. etc., and there really is a yellowish orange patch on the wall.' Also, when I have a yellow red afterimage, 'a vast mechanism ... expresses a temptation (behaviour disposition) to say, "There is a yellowish red patch on the wall."' But having an orange afterimage isn't a bit like seeing an orange coloured patch on the wall. *How* are the two supposed to be alike? Here are some of the differences. (1) The colour of the patch is in the surface mode. (2) The afterimage colour lacks texture and grain. (3) Afterimages are relatively unstable and change in colour. (4) Afterimages lack sharp outlines. (5) Afterimages move with the eye. (6) Afterimages are in some sense self-illuminating, and they can be seen in the dark. But the patch is not self-illuminating. And so on.

And I have no temptation to say, when having an

[23] (London: Routledge, 1963), p. 94.

afterimage *with my eyes closed*, 'There is a yellowish red patch on the wall.' It is true that under certain special circumstances afterimages (or the experience of having them) can be mistaken for discolorations on surfaces, or, say, for traffic signals at night – red for green – but this has no tendency to show that the two kinds of experience are alike in some unspecified respect. And if the respect in which they *are* alike – colour – is specified, then it has been conceded that afterimages do after all have colour.

6.7 The Failure of Physicalism

The example of coloured shadows, however, does suggest that there is a connection between the Wilson and Brocklebank property for physical colours and for physiological colours. And this is in fact the case. A coloured shadow appears when two light sources give shadows of an object. Let one light be stronger than the other, and let it be coloured red. Let the weaker light be colourless. The total illumination is reddish white. In this reddish white light a shadow or partial shadow *which does not change the colour of the illumination* is a 'neutral' or 'grey' shadow. So a shadow from which the red component has been removed is neutral minus red, or *less red* than neutral. This is green. So the shadow from the red light is a green *darkening* in this light. Here adaptation of the sort involved in the perception of external objects is clearly at work. In this sense there is a connection between what is going on in me (adaptation) when I see a coloured shadow or an afterimage and when I see a real orange patch on a wall in front of me. What is needed here is not a reduction describing what happens at the surface of perceived objects (afterimages have no surfaces) and the physics of this, but a wider understanding of the whole relationship between the organism, the eyes and the visible scene, the physiology of colour vision, and ultimately the evolution of the colour sense. Where something as complicated as colour perception is not understood, it is surely more sensible to seek to develop a theory which does provide understanding, rather than to attempt to reduce the puzzling item to something else which is simple enough to be easily understood. I should say that I have no objection in principle to reductions where the theory to be reduced is in good order, whether physicalist, spiritualist, or any other kind – except that they never seem to work. In fields where there is still much to be learned, as in the physiology and psychology of colour perception, what is needed is not a reduction which so to speak polarizes our knowledge into the primary and physical, and the secondary or derivative or illusory, but clearer and more imaginative theory. The trouble with physicalism is that it takes a still distant scientific ideal (the unity of knowledge) as a present fact. The ideal is not wrong, but we have not yet achieved it, and materialism is a premature synthesis. A lively, growing science such as psychology must have available more flexible and many-sided concepts than can be given by any more formal system of concepts.

Armstrong has mistaken a certain type of specification for an essence. His claim that 'the scientific image of the world has to be taken *seriously*. It has to be taken ontologically'[24] is a false equation. It is as if Armstrong, in giving his solution to the problem of length universals in *A Theory of Universals* II, had glanced at the colours of the spectrum (missing white, brown, grey, black and purple), neatly lined up against wavelength values, and taken the very superficial analogy seriously in his sense – ontologically. Why should every scientific specification correspond to some clanking great disjunctive essence physicalistically conceived? Colour science has for a long time interested itself in the departure of hue from wavelength – without becoming *frivolous*. These variations do not mirror the physical as it is described in physics, but all this means is that the world of the physicalist is not complete. The physicalist specification is not ontological, somehow more serious and *deeper*,[25] and what it leaves out illusion or mere appearance. It is more superficial, and the image of depth which it excites is an illusion.

[24] Armstrong, *A Materialist Theory of the Mind*, p. 271.
[25] Armstrong, *Universals and Scientific Realism*, p. 126.

J.J.C. Smart has come to accept a physicalist view of colour similar to Armstrong's, according to which 'a colour is a physical state of the surface of an object, that state which normally explains certain patterns of discriminatory reactions of normal human percipients.'[26] We have seen that this view is false. Yet what alternative is there for physicalism? If physicalism cannot give a coherent account of colour, then it cannot be true. Opponent process theory, as this has been set out for philosophical purposes in a series of papers and in a recent book by Larry Hardin,[27] would seem to offer the last hope for a physicalist theory of colour. Yet it faces insuperable difficulties. Hardin's theory implies that the non-existence of red-green is a contingent matter, and this seems just as wrong-headed as the idea that the non-existence of a whole number lying between 3 and 4 is contingent, and not for very different reasons. As Russell observed in his criticism of Kant's theory of necessity,

> Moreover we only push one stage farther back the region of 'mere fact', for the constitution of our minds remains still a mere fact. The theory of necessity urged by Kant ... appears radically vicious. Everything is in a sense a mere fact.[28]

Furthermore, Hardin's account involves what he himself calls a complete 'subjectivism' about colours, and this cannot be expected to appeal to robust Australian realists – whatever it means.[29]

[26] In 'On Some Criticisms of a Physicalist Theory of Colours', in C.Y. Cheng (ed.), *Philosophical Aspects of the Mind-Body Problem*.

[27] See in particular C.L. Hardin, 'A New Look at Color', *American Philosophical Quarterly*, 21, 2 (1984), pp. 125–133, and *Color for Philosophers* (Indianapolis: Hackett, 1986).

[28] Bertrand Russell, *The Principles of Mathematics* (London: Allen & Unwin, 1956), 430, p. 454.

[29] C.L. Hardin, 'Are Scientific Objects Colored?', *Mind*, xcii (1984), pp. 491–500.

7

Impossible Colours and the Interpretation of Colour Space

> Has nature herself nothing to say here? Indeed she has – but she makes herself audible in another way.
>
> WITTGENSTEIN

7.1 Two Types of Colour Concept

Very little has been said so far about *colours*. The main question of Chapter 2 was not why the colour white cannot be transparent, but why white coloured surfaces or objects cannot. Chapter 4 defines a grey area but not the colour grey. The treatment of brown in Chapter 3 is an exception, because the definition includes another substantive colour term ('yellow'), and this made it not only possible but necessary to define the *colour* brown. Chapter 5 is partly about blackness, which is not a colour. The colour incompatibility problem of Chapter 6 concerns the impossibility of some body or thing being red (all over) and green (all over) at the same time. It is the problem of predicate incompatibility. This should be compared with the quite different question which Wittgenstein asks in *Remarks on Colour*: whether there can be a reddish green. (Note the spatial and temporal references and the reference to a body in the first question.) We cannot ask why there cannot be a reddish green all over at the same time. This is meaningless. Wittgenstein's new problem could be called the problem of subject incompatibility.[1]

Unlike the old question, the new one does not apply to

[1] The problem is not new in the history of colour science. It is the basis of Hering's Theory and was discussed by Helmholtz, who claimed to be able to *see* yellow as a reddish green. What is new is the philosophical setting in which Wittgenstein puts the question.

all combinations of colours. Blue can be greenish, but red cannot. And conversely a body which is green (all over) can no more be blue (all over) at the same time than it can be red (all over) – or anyway, not the same blue.

Wittgenstein opens *Remarks on Colour* I with a group of observations distinguishing propositions referring to colours as such from propositions referring to the being coloured of coloured bodies, or what we might call their colourednesses.

> A language-game: Report whether a certain body is lighter or darker than another. – But now there's a related one: State the relationship between the lightnesses of certain shades of colour. (Compare with this: Determining the relationship between the lengths of two sticks and the relationship between two numbers.) The form of the propositions in both language-games is the same: '*X* is lighter than *Y*'. But in the first it is an external relation and the proposition is temporal, in the second it is an internal relation and the proposition is timeless.[2]

There is evidently no easy logical passage from propositions of the first type to propositions of the second type. Much philosophical writing about colour is vitiated by a failure to observe Wittgenstein's distinction, and is disfigured by confusions between colours and coloured patches or coloured areas, colours and coloured things, and other similar mistakes.

In *Individuals*,[3] for example, Strawson claims that colours, unlike sounds, are 'intrinsically spatial'. But as he enlarges on the significance of this, it emerges that he is discussing a 'visual scene', in which we are presented with 'coloured *areas*' (my italics) which are spatial but not colours. The scene is then broken up into its 'uniform elements', which are *patches* of determinate hue, saturation and brightness. These patches are then said to be above, below, to the right of, and to the left of one another. The visual elements are finally said to be 'the momentary states of the colour-patches of the visual scene' – note the

[2] *Remarks on Colour*, I 1.
[3] (London: Methuen, 1959), pp. 78–80.

hyphen. It could just as easily be shown that numbers are intrinsically spatial, or that numbered areas and the momentary number states of the uniform elements of the countable manifold are.

Failure to observe the distinction is also responsible for the (in my view) confused ontology of colour-particulars defended by G.F. Stout.[4] Ignoring the English language, Stout argued that there are as many colours as there are distinct uniform patches of colour, and that colours are 'as particular as the places they are in'. Helen Knight has charged Stout with confusing colour and place, and in this she is surely correct.[5] Stout writes that, 'A colour . . . is co-extensive with the place in which it is.' If a place is defined independently of colour by its polar co-ordinates, then clearly the (patch of) colour and the place need not be co-extensive. If on the other hand they are not defined independently, then colour and place are confused. A colour 'is spread out over the surface so that the larger the coloured surface the more colour there is' (he should have written, 'the more colours there are'), but also, inconsistently, that, 'my point is that the same colour cannot at the same time be in two places'. If a bloodstain seeps out over a place in the visual field previously occupied by a white rug, then it occupies places previously 'covered' by the rug or its colour. But the colour of the blood cannot, according to Stout, be in its original place and in the places vacated by the rug at the same time, unless 'it' changes colour as it travels. The colour is so to speak trapped in its original place, because it is co-extensive with it. Or if it carries its place with it, becoming a larger element and occupying a larger place in the visual field, obviously colour and place are confused, in the sense that a place is not defined independently of colour. And Stout pretty well concedes the point: 'In visual space coloured extension is simply identical with extended colour.'

Russell writes in the *Inquiry into Meaning and Truth*,[6]

[4] G.F. Stout, 'Universals Again', *Proceedings of the Aristotelian Society*, Suppl. Vol. **15** (1936), p. 9.
[5] 'Stout on Universals', *Mind*, xlv, 77 (1936), pp. 54–55.
[6] (London: Allen & Unwin, 1950), Ch. 6, 'Proper Names'.

that the angular coordinates (θ,ϕ) of a place in the visual field 'may be regarded as qualities'. If C is a colour, then according to Russell we may say that a 'thing' is a bundle of qualities (C,θ,ϕ), and this thing is at the place (θ,ϕ), 'and it is analytic that it is not at the place (θ,ϕ)'. So 'things' are incapable of movement, which is absurd. Here 'things' are confused with the places which they occupy, and the analyticity of the immobility of things derives from immobility of the places with which they are confused.

G.E. Moore deduces what could be called the bare identity view (that two things can differ *solo numero*) with the aid of the false premise that colours admit of spatial predicates.[7] 'We can never say, "The red I mean is the one surrounded by yellow, and not the one surrounded by blue." For the one surrounded by yellow is also surrounded by blue; they are not two but one, and whatever is true of that which is surrounded by yellow is also true of that which is surrounded by blue.' Moore takes this to be a reductio ad absurdum of the negation of the bare identity view, or of the view 'that there is no difference but conceptual difference'. The application of Leibniz's Law is correct, but the conclusion Moore ought to have drawn is that the *colours* he mentions are not surrounded by other colours. The red *patch* is surrounded by a yellow patch, but this red patch is *not* also surrounded by a blue patch. The coloured patches are two, admit of spatial predicates, and are surrounded by other patches. What the application of Leibniz's Law really shows is that these truths cannot be transferred to colours.

An appreciation of the full significance of this point and the special way in which colour (and not just colours) is non-spatial would, however, require a close comparison of the concept of a colour and colour with other parallel pairs of concepts – for example, those of shapes and shape, weights and weight, sounds and sound, smells and smell, and so on, where similar principles might or might not obtain. What it does suggest, pending such comparisons, is that, for whatever reason, Wittgenstein's atemporal and

[7] 'Identity', *Proceedings of the Aristotelian Society*, 1900–1901, p. 111.

internal propositions about colours cannot be reduced to temporal external propositions. From 'Brown is unsaturated', for example, we cannot infer that 'Brown objects are unsaturated.' This is nonsense. Saturation is a dimension in which colours, but not objects, can be scaled. Nor can we infer that yellow is a bright colour from the fact that yellow objects are. Even if all yellow objects are bright, this is not sufficient. If all dark blue objects are destroyed, leaving only light blue ones, this would not show that blue is a light colour. What would have to be shown is that the remaining blue objects were light because they had to be, that there was in fact a necessity here. We would come back to the fact that blue is necessarily not a light colour. Wittgenstein's necessary truths about colours (e.g. that saturated yellow is lighter than saturated blue)[8] cannot be inferred from contingent truths about the colours bodies are coloured. And in general the attempt to reduce the terms 'saturation', 'brightness' and 'hue' to the terms of physics is, as we saw in Chapter 6, a failure.[9] We cannot, therefore, transfer the analysis of predicate incompatibility to the new problem of subject incompatibility. It would in any case show too much. The solution to the problem of subject incompatibility must show that reddish green is impossible, but not that bluish green is.

Yet there is a connection between colours and coloured things. There is an obvious type of argument which proceeds from the truth that nothing can be red (all over) and

[8] Remarks on Colour, III 9.

[9] In Chapter 6 I avoided mention of David Armstrong's confusing belief that 'is red' is not a predicate but 'red' is, that 'there are such predicates as 'coloured' and 'red' but there is no property *being coloured* or *being red.*' (*Universals and Scientific Realism* II, p. 117.) Armstrong's argument for this conclusion was neatly refuted by David Sanford in 'Armstrong's Theory of Universals', *British Journal for the Philosophy of Science* **31** (1980), p. 76. For me as for Frege (Michael Dummett, 'Frege's Philosophy', in *Truth and Other Enigmas* (London: Duckworth, 1978), p. 99, 'is red', in the sense of 'is coloured red', is the predicate, *being coloured* or *being red coloured* is the property, and the colour red an entirely different kind of object: a logical subject. If my criticisms of the physicalist theory of colours are correct, then it will appear to Armstrong as if it has been shown that colours do not reduce to light emissions of different wavelengths, but from my point of view it will merely have been shown that Armstrong's analysis of *being coloured* is wrong.

green (all over) at the same time to the truth that there cannot be a reddish green, and very nearly succeeds.

Assume	(1)	If red-green is a possible colour, then something can be coloured red-green.
	(2)	If something can be coloured red-green, then it can be coloured red (all over) and green (all over) at the same time.
	(3)	Nothing can be coloured red (all over) and green (all over) at the same time.
Then	(4)	Nothing can be coloured red-green (2,3).
So	(5)	Red-green is not a possible colour (1,4).
But	(6)	If reddish green is a possible colour then red-green is.
So	(7)	Reddish green is not a possible colour (5,6).

This argument establishes the right conclusion, and it is obviously valid. But it is not sound. (2) is clearly false. From the fact that something is blue-green coloured it does not follow that it is or can be blue (all over) and green (all over) in the sense intended in the literature on predicate incompatibility. Moreover, the argument offers no positive or general insight into the relation between colours and coloured things. What is needed is a principle which extends the type of polarity used in Chapter 6 to explain incompatibility at the physical level to the incompatibility of colours as such.

7.2 Varieties of Colour Space

Wittgenstein's new problem of subject incompatibility invites an answer based on the geometrical arrangement of colours in the colour circle. He had already referred to a geometry of colours in *Zettel* ('I want to say that there is a geometrical gap, not a physical one, between red and green.')[10] and he considers the idea of a geometry of colour space again in *Remarks on Colour*.[11] But is it more than a

[10] L. Wittgenstein, *Zettel*, ed. G.E.M. Anscombe and G.H. von Wright (Oxford: Blackwell, 1967), p. 65e.
[11] E.g. at *Remarks on Colour*, I 66.

metaphor? The fact that the chemical elements can be arranged in the logical or geometrical order of the periodic table does not mean that the gap between gold and silver, for example, is geometrical and not physical. What would this mean? The fact that the most effective means of representing the order of elements is geometrical does not mean that the basis of the order is a geometrical one.[12] What is wanted is a fuller understanding of what the colour circle or hue circuit *is*, which would consist of an explanation of its origin or basis, just as with the periodic table. Is there a theory which would generate the arrangement of all the colours in an order, of which the hue circuit would be a cross section?

There are in fact many models of colour space. Höfler's model is an octahedron, with red, yellow, blue, green, black and white at the vertices, and there is a similar model from Ebbinghaus. Chevreul's system is a hemisphere, and Titchener's model deforms the tetrahedron, so that yellow is closer to white than it is to black. There is a sense in which the colour sphere, deriving from Runge, places black and white in the same dimension as the chromatic colours. Ostwald's model is a flattened cone or top. The Munsell solid is an irregular tree. The Inter-Society Colour Council National Bureau of Standards system is a refinement of Munsell's. The newer Swedish Natural Colour System includes the unitary hues as an essential feature. There are also complex solids whose inspiration derives as much from mathematics as from empirical colour science. The term 'colour space' is most commonly used to refer to the two-dimensional chromaticity diagram, a descendant of Maxwell's triangle, which gives a summary of the facts of light mixing at fixed brightness.

It must be clearly understood that these systems and models have entirely different functions and purposes, and they record different types of data. They are also governed by different types of concepts, some physical, some

[12] Or does it? Is it the same to say: There is a geometrical gap between Ag and Au? Cf. 'There can be a nor'-wester but not a nor'-souther' and *Remarks on Colour*, I 21.

psychological, and some a mixture.[13] Ostwald's cone place complementary colours opposite, so that unitary red is no opposite unitary green. This arrangement also has th effect of making the spacing between the hues unequal, s that on the cool side of the solid there are too man blue-greens which are more similar to one another than th warm colours which lie opposite them. But there is n absolute requirement of equal spacing, nor is there an absolute reason that complementaries should fall opposit one another, except that this arrangement would facilitat the prediction of the results of additive mixing by Newton' barycentric method. Similarly, the only point of the fun damentally free-hand octahedron and the colour square i to give special emphasis to the unitary hues. (It is interest ing to see Wittgenstein reviving this type of model,[14] in par with a motivation similar to that of the phenomenologists as one type of fact 'that is independent of both the stimulu and psychology').[15] In a recent article on 'Compromises i Colour Solids', Joy Luke Turner writes,

[13] There is a complete muddle in some discussions of contemporary colou theory concerning the relation of the chromaticity diagram (C1) to the psycholo ical colour solid (C2). (C1) is supposed to represent the stimulus, or even *be* th stimulus, and (C2) is much richer than (C1). The difference between the stimul model and what we see is to be given by physiological functions. But of cour (C1) is not even a model of the stimulus. It is a record of the results given b observers who are instructed to match lights of different spectral compositio under restricted viewing conditions. And (C2) is not 'the response' or a model it. It is a record of observers' judgements of relative similarity.

[14] In *Wittgenstein's Lectures 1930–1932*, ed. Desmond Lee (Oxford: Blackwe 1980), p. 8 and p. 11.

[15] E.G. Boring writes in *Sensation and Perception in the History of E perimental Psychology* (New York: Appleton, 1942), p. 149, 'For a while Ebbin haus' double pyramid represented the last stand of the phenomenologists again the encroachments of the nervous system upon psychology: here in the colo pyramid, it was argued, there is at least one fact which is independent of both th stimulus and psychology. That there are but few psychologists any longer t cherish such a last leaf on the tree of mentalism goes to show how phenomeno ogy perpetually in the development of psychology, loses the battle to e perimentalism – until Wittgenstein redrew the boundaries of psychology and log in such a way that this type of fact *could* not be poached by 'experimentalism'. I the late twenties and thirties Wittgenstein was apparently seeking 'a pure phenomenological theory of colours' which would include nothing 'hypothetica such as light waves or the physiology of the retina' (Rush Rhees, 'The Tractatu Seeds of Some Misunderstandings', *Philosophical Review*, lxii (1963), p. 217. Th search continued in *Remarks on Colour*, but Wittgenstein was now clear that wh

> If the interest of the user [of the colour solid] is in industrial color [sic] control, perceptively equal small color differences are important and it is also important that the color notation can be derived in as simple a way as possible from the instrumentally read colorimetric data. It is less important to these users that complementary hues fall opposite one another in color space.[16]

After making some further related points about the requirements of use, she concludes,

> The point of all this is that color solids represent a set of compromises made in many cases to achieve a tool. They can contain large errors from one viewpoint and be very practical from another. The catch is that the user should be aware of what the compromises are. I firmly believe in a continuous effort to develop a color solid which represents as accurately as possible the human visual experience, but I don't believe that means that specialist solids have to be discarded.[17]

For what I propose to say about colour incompatibility, however, we can consider just the standard idealized double cone with the three dimensions of saturation, hue and brightness. It is a significant fact that the overall order flows from one principle alone. Colours are to be placed near to one another according to relative similarity. (The double cone could therefore be considered as a set of propositions or statements.) Two very similar yellows will be placed very near to one another, two identical yellows will be placed on top of one another, so that no sample appears twice, and two very different colours, such as yellow and blue, will end up far apart. This order is commonly said to be an order of colours, but this cannot be strictly correct. The usage will produce the familiar but ridiculous claim that there are ten or seven and a half or however many million *colours*. We should not of course say in the same breath, or

---

was wanted was 'not a theory of colour (neither a physiological one nor a psychological one), but rather the logic of colour concepts. And this accomplishes what people have often unjustly expected of a theory' (*Remarks on Colour*, I 22).

[16] Joy Luke Turner, 'Compromises in Color Solids', *Color Research and Application* **5**, 3 (1980).

[17] Ibid.

at all, that red is a colour and that two just noticeably different shades of red are different colours. Then the question would arise whether either or both of them was the same colour as red, or a different colour. A colour in the most central sense, the sense in which blue, brown, black, pink, yellow and violet are colours, is a *volume* in the three-dimensional space, not a point. The facts which the space represents, for example the fact that it does have volumes, the opponency or polarity of the blue-yellow, red-green and light-dark axes, the circling character of the cross section which is the hue circuit, and so on, can be regarded as calling for an explanation just as the order of the periodic table does. It is tacitly agreed on all sides that the similarity colour space, as I shall call it,[18] does express the nature of colours or what colours in their nature are. Thus for Goethe it would illustrate the 'suffer' and 'do' of light, the primary polarity of yellow and blue, the principles of polarity and Steigerung (augmentation, culmination). For the modern followers of Hering, these polarities are manifestations of the basic opponent principles of the visual system, with three opponent channels. For the physicalist the similarity space would be a transformation of part of the electromagnetic spectrum, and it would represent differences of frequency. For Wittgenstein, the ordering of colours is grammar in his extended sense, and it could be regarded as a visible expression of our colour concepts. We can, however, take the similarity space phenomenologically, and consider the incompatibilities and other relationships which it illustrates and fixes as manifestations of the deeper order without choosing between these alternatives. I suspect, however, that, at the end of the day, the first, second and fourth alternatives are all going to prove compatible, both with one another and with the account of being coloured given in Chapters 2–6.

[18] I prefer to avoid the phrase 'psychological colour space', as there is no clear argument which shows why or in what sense the order is psychological. This is, I suspect, a muddled conception which has no more basis than the idea that the periodic table is a partly psychological order because it results partly from paying attention to the appearance of elements, or the idea that a sequence of geometrical shapes must be a psychological order because it is arranged by *people*.

7.3 Two into One Won't Go

Consider now, by way of comparison with the colour incompatibility case, two subjects that are not only different subjects but different *kinds* of subject. Can we ask why something cannot be a mouse and a chainsaw at the same time? Are the two things compatible, in the sense that some one thing could be both? Someone who seriously put these questions would have a difficult time making a case for the claim that he had anything but the loosest grip on what the two very different things *are*. Where, for example, will the legs of the new hybrid creature go? What will its bones be made of? Where will the eyes go? Wherever the eyes go, we have mouse manifestations, not chainsaw manifestations. Will the mouse/chainsaw be a timorous animal? Will it have a psychology? Will it ever *panic*? Will it cut down trees? How? Will it feed or be filled up? What sort of physiology/motor will it have? Will it have a reproductive system? Such difficulties would of course be less if the incompatible things under consideration were both organisms, say a mouse and a shark, or a butterfly and an eel. But they would be of essentially the same type. With an organism and a mechanical tool, the misguidedness of the question is just that much clearer. For a mouse is, essentially, a living being, and a chainsaw is not – in spite of Putnam's claim that cats might turn out to have been robots controlled from a nearby possible world. Anything that was both a mouse and a chainsaw would have to be both a living being and not a living being, which is logically impossible. One has only to bring to mind the extraordinary fineness of the concepts involved in our zoological schemes of classification and their evolutionary basis to appreciate the havoc which would be caused in these schemes by the insertion of fictitious items constructed without reference to the principles ordering the things which are being classified. Something similar would be true of the advertising catalogues of chainsaw manufacturers.

What is more difficult in the case of colours is to find a means of bringing out the disruption in our concepts or schemes of classification which would result from inserting a fictitious colour into colour space. The difference, however, is only due to the fact that the mouse and the chainsaw have more and more various properties than the two incompatible colours. We must be on our guard against specious arguments from imaginability. It must be insisted that words and images purporting to capture new *possibilities*, but not based on the order which *gives* possibility in the domain, are incoherent. An impossible colour is, to appropriate a phrase from Leibniz, *être de raison non raisonnante*, just as much as an impossible but apparently convincing object in any other real domain. What is a little harder in the case of colours is to spell out the meaning of the *non raisonnante*. An interesting intermediate case would be the incompatibility of two metals, such as silver and gold. Here the two incompatible subjects are of the same kind (chemical elements) but the incompatibility is presumably at bottom no less demonstrably logical than with incompatible objects of different kinds. What must generate the incompatibility is the chemical natures of the two elements. These can presumably be stated in such a way as to make the contradiction apparent. It is possible, however, without a reduction of this type, to show that the insertion of a fictitious colour into our three-dimensional colour space will disrupt the order and prevent us from conceiving some other colour or group of colours in the space, independently of the explanation of this fact in the generative basis of the space. For colours and the similarity colour space are inseparable. The positions of the colours on the hue circuit, for example, are determined by the positions of their intermediaries and vice versa, and these together determine the geometry of the space.

Suppose we choose a location for reddish green in colour space. Some location must be chosen, for (a Wittgensteinian point) if the colour is located *outside* colour space, we will have deprived ourselves of the means of making good the claim that it is a *colour* which is being imagined. A colour located outside colour space would have no fixed location or address. It would be of no definite hue, of no particular saturation, and of indefinite brightness. Obviously reddish green must be located somewhere between *red* and *green*. It must not, however, be located exactly between them at the centre of the space, (i) because this address is already occupied by grey, and (ii) because reddish green must be capable of being nearly as saturated as its parent colours. So let us place reddish green between yellow and green among the saturated yellow-greens. This choice is arbitrary, but exactly the same results are obtained with any location. (In fact the difficulty is precisely that any location is arbitrary.) Now it becomes true: (1) that a certain yellowish green is not intermediate to yellow and green. It has been displaced by the reddish green and it must be relocated elsewhere, thereby displacing another colour, which . . . ; (2) that not all shades between yellow and green are yellowish or greenish – there is one which is reddish; (3) that not all reddish hues are continuous with unitary red; (4) that not all mixed or binary hues can be desaturated and retain their position in the dimension of hue, because unsaturated reddish green becomes saturated yellowish green; and so on.

It might be thought than we can after all imagine the new colour in the empty space beyond the hue circuit, say between yellow and green and with the intermediate hues filled in. We extend the colour space. But this too contradicts the principles on which the space is organized. The new edge of colour space on which the saturated reddish greens appear is now the new limit of saturation. This means that the most saturated yellowish greens *are not maximally saturated*. And a maximally saturated yellowish green is not yellowish but *reddish*. Imagining the false colour in this location means ignoring the information which is represented in the rest of the similarity colour space. The idea of reddish green in the chosen location is actually incompatible with a consistent application of the concept of saturation.

And if the principles on which similarity colour space is constructed are ignored, so that reddish green can be located anywhere we choose, why not locate it next to bluish orange, another impossible colour, however that is located? We can produce an infinity of verbally describable but incoherent colours by this means, and the number of lost truths will multiply to unmanageable proportions. Can this question be answered: if reddish green is a possible colour and therefore an actual location in colour space, what location or colour *can't* there be, and why? Could there be a reddish greenish bluish blackish whitish pinkish *magenta*? Or a lilac green? What prevents these compound terms from getting a grip on our imagination is not just the number of terms they contain, but as Wittgenstein saw, our powers of imagining or conceiving *are* the representations of colours in colour space, or colour itself. What is offered by some of the elements in the spurious compound terms – a location in colour space – is withdrawn by the others, so that, appearances to the contrary, what is given is not a possible representation or a clear set of directions – any set of directions – for finding the place in colour space.

7.4 The Contradiction

This is probably as close as we can get to anything resembling a contradiction in the idea of an impossible colour, considered as part of the order of colours. It is, however, close enough to allow a very hard interpretation to the 'cannot' in 'There cannot be a reddish green'. Obtaining a fully logical contradiction would depend on agreement on the interpretation of the similarity colour space ('suffer' and 'do' of light, neural coding, frequency order, logical grammar), including its exact relations to the physics, physiology, and psychology of colour; or, what is the same thing, the reduction of colour space to colour perception and its proper objects. The fact is not that there is no easy way to find a contradiction, but that there are too many. What this suggests is that the various different theories have different areas of competence, physical, physiological, psychological and logical. The existence of a purely qualitative similarity space suggests that there is a harmony between the four fields which will resist any sort of one-sided interpretation.

8

Simplicity

> All our forms of speech are taken from ordinary physical language and cannot be used in epistemology or phenomenology without casting a distorting light on their objects.
>
> WITTGENSTEIN, *Philosophical Remarks*

> Things which are uniform, containing no variety, are always mere abstractions.
>
> LEIBNIZ, *New Essays on Human Understanding*

There is a simple connection between the idea that colours are spatial and the idea that they are simple. It is this. If colours are spatial, then they are simple. In this chapter I shall try to show that they are not simple. I shall claim that philosophers who have believed in their simplicity have confused logical simplicity with visual uniformity. They have confused the kind of simplicity which could intelligibly be attributed to coloured patches or expanses of colour with the kind of simplicity which could be attributed to colours as such. So part of my purpose in this chapter is to disentangle our concepts of the colours from spatial and material images which represent them.

The first question that must be asked is what the simplicity in question is, and how the complexity which philosophers have wished to deny to colours is to be understood. The doctrine is notoriously not simple or clear in either Locke or Hume. I shall have something to say about them later. For the moment, here are three examples of what more recent philosophers have had to say:

> Words like 'red' and 'green' cannot be unpacked in the way that words like 'uncle' can; since they are introduced

ostensively they cannot be resolved into conjunctions of simpler concepts comparable with 'male sibling of a parent'.[1]

> A man sips two kinds of port and notices a difference between the two flavours. He may learn names for them, and assent to the proposition that they cannot both be in the same sip, and yet he may be quite unable to expand his two designatory rules verbally ...
>
> The colours, like the flavours, contain no separately noticeable marks. But this simplicity is more unexpected in the case of colours because sight is a successful organizing sense, and most words for seen qualities find their places in elaborate logical networks. Also colours, unlike flavours, can be pointed to, and one is surprised by the discovery that there is a limit beyond which pointing cannot select from within a quality further marks of that quality.[2]

> I ask how the generic character P (being red) is supposed to be related to the specific characters R_1, R_2, etc. (different shades of red). P is a colour quality contained within each specific red ... Each specific red would have two constituents, distinguishable though inseparable. There would be on the one hand the generic quality, and on the other its specification. My difficulty is that I cannot by any kind of scrutiny discern any such complexity in any specific red.[3]

These different versions of the doctrine raise rather different questions. I regard it as a *dogma*, because it is based on the claim, unargued, that the concepts of colours cannot be 'unpacked', that there are no noticeable features or marks by which the *qualia* can be identified, or that we can all just look and *see* that they are not complex. (Locke's justification of the dogma at *Essay* II, ii, 1 is interestingly no argument but straight assertion.) Remnant says that *since* colour words are introduced ostensively, they cannot be resolved into simple concepts. Does this follow? The

[1] Peter Remnant, 'Red and Green All Over Again', *Analysis* March (1961), p. 93.

[2] D.F. Pears, 'Incompatibilities of Colours', in A. Flew, ed., *Logic and Language*, 2nd ser. (Oxford: Blackwell, 1966), p. 119.

[3] G.F. Stout, 'Universals Again', *Proceedings of the Aristotelian Society*, Suppl. Vol. 15 (1936), p. 14.

missing major premise needed to make the enthymeme deductively valid will be to the effect that whatever word is introduced ostensively cannot be unpacked. This is clearly false. To many people the words 'tree' and 'horse' are introduced ostensively, and very probably 'uncle' and 'triangle' as well. But this does not mean that these words cannot be unpacked. They can. It would only mean that if they were introduced ostensively because they *had* to be, because there was no other way of introducing them, and this in turn would be because the words could not be unpacked. But whether or not they can is exactly the question at issue. Pears protects the dogma by creating a necessary truth here. 'Few words are given full definitions in dictionaries, but "red" and "green" could not be, since they are words which are necessarily taught almost entirely by examples.'[4] Stout invites us to *scrutinize* a colour, and to see if we can detect the two constituents which on his understanding of how, for example, scarlet is related to red would have to be there. The point he makes is that he fails to see scarlet *and* red. (Do we see an oak *and* a tree? Are they (?) visually 'distinguishable though inseparable'?) Stout supposes that the two constituents would have to be *visually* distinguishable. We are to scrutinize a colour and literally *see* that it has no constituents. (Nothing happens either, though, when we scrutinize uncles. They do not reveal to visual scrutiny two constituents.)

And how does one scrutinize a colour? What would the difference be between scrutinizing the colour and the coloured area or patch of colour? In fact in the same article in which Stout confesses his phenomenological difficulty he also argues that a colour is identical with a coloured area, that 'in visual space coloured extension is simply identical with extended colour'.[5] Overwhelmed by the uniformity of the coloured area, we may not stop to argue. And so the dogma is established, but only for colours as Stout means them, areas or patches of colour as we would call them, and not for unextended things like colours.

[4] Incompatibilities of Colours', p. 112 n. 1.
[5] Stout, 'Universals Again', p. 11.

The Humean challenge, as Bernard Harrison called it in *Form and Content*,[6] is '*Show* me how the colours are to be described'. This is the real issue. But the inability to show this by itself shows nothing. In particular it does not entitle us to claim that there *is* no way in which colours can be described. For the inability is actually not the issue. Pears's port-sipper is not only unable, Pears says, to expand his designatory rules; after all, perhaps he hasn't given the matter much thought, he is no professional wine-taster, he is feeling sleepy after the ports, and the point is not his inability but a much stronger concealed claim – he is *quite* unable, says Pears. How does he know? The claim is being advanced as an empirical report about the port-sipper's abilities, but it is in fact a strikingly a priori one. What Pears is suggesting is not only that he is unable to expand his rules for the application of the words but (a much stronger claim) that he is unable to do so because there *is* no expansion. But why couldn't the port-sipper have said that the first sip was sweeter than the second, that the second was more salty and sour, though less bitter, and so on? Moreover, 'port' is already the name of a very specific kind of taste as well as a wine (cf. 'orange', a colour and a fruit) which carries the description (*OED*), 'a strong, sweet and slightly astringent taste'. Why should one not say of an as yet undescribed taste that it was a *port* taste? The marks of the taste are that it is port, strong, sweet and slightly astringent, and specific ports will be marked off from one another in the standard dimensions of taste (sweet, bitter, sour and salt). The port-sipper is confronted with two *very similar* tastes. Would the discussion proceed as smoothly with the port-sipper tasting port and claret, or port and beer, or port and honey, or port and a rotten egg? Some less regimented marks of tastes: insipid, mild, spicy, foul, brackish, savoury, coarse, rank, luscious, tart, and so on.

Pears does not say what he understands by the term

[6] Bernard Harrison, *Form and Content*, p. 9. Harrison's actual words are, '*Show* me how the raw feels are to be described.' A colour cannot be a raw feel because it implies a division of colour data into colours, but the point remains the same.

'mark', or how it differs from 'feature', 'characteristic', 'aspect', etc. I shall mean by the mark of a colour or a mark of a colour a property of the colour which either by itself or together with other marks serves to identify the colour. A mark of a colour will (*OED*) separate it from other colours 'as by a line or distinctive mark'. My use of 'mark', suggested by Leibniz's use of *nota* in the 'Meditations on Knowledge, Truth and Ideas',[7] needs to be brought into some definite relation with Frege's very strict use. For Frege G is a mark of a first-level concept F if and only if anything that falls under F falls under G. So G is itself a first-level concept. As I use 'mark', *chromatic* is a mark of red, but it does not follow that every body that is red is chromatic, or that any body is chromatic; this is nonsense. To understand the concinnity of my usage and Frege's, we must respect the ambiguity of 'is red'. It can mean 'is a shade of red', as in 'This (shade) is red or a red', 'is coloured red', as in 'This (book) is red', or 'is the same thing as the colour red', as in 'This (colour) is red'.[8] Any shade that is a shade of red is chromatic; any colour that a body is coloured is chromatic if it is red; if anything is the same thing as the colour red, it is chromatic.

The makers of dictionaries do define colours, but only ostensively, or as it were by mediate ostension. They name *examples* of e.g. red-coloured things – ripe apples, sunsets, strawberries, etc. – although sometimes they do appear to be attempting to supply the marks or specific differences: 'red is a colour which is found at the longwave end of the spectrum'. This also turns out to be an example of something coloured red (the longwave end of the spectrum), even if the spectrum is Rays and the Rays to speak strictly are not coloured. But it will take an argument to show that this is *not* in the above sense a mark of red. And this is not

[7] Leibniz, *Philosophical Papers and Letters*, ed. Leroy E. Loemker (Dordrecht: Reidel, 1976), pp. 291–292.

[8] 'The word "red" used as a noun is a proper name and must stand for an object; a predicate like ". . . is red" is an expression of such a totally different kind that we cannot suppose it to be correlated with an entity of the same sort at all' (Michael Dummett, 'Frege's Philosophy', *Truth and Other Enigmas* (London: Duckworth, 1978), p. 100).

the only kind of mark available. How can it be, if the dogma is correct, that the *OED* definition of 'brown' ('gloomy, dark, serious, name of a composite colour produced by a mixture of red, yellow and black') is actually *inferior* in some respects to Webster's 'any group of colours between red and yellow in hue, of medium to low lightness, and of moderate to low saturation'? Webster has surely unpacked the word for brown. If this isn't unpacking the word for brown, then nothing could be. Moreover, the complete definitions of the colours, or rather of the being coloured of coloured objects and materials, proposed in Chapter 6 and illustrated in Fig. VIII, could easily be expanded into words, as they must be if we are to bring the contradiction to light in the proposition that something is or could be red and green coloured all over simultaneously.

The Webster definition of brown gives us an idea of brown, but not in the equivocal Lockean sense; it gives us the concept of brown. If the claim of the friends of simple ideas of colour is that this definition will not give someone unacquainted with the colour the *concept* of the colour, then what they claim is just false. The concept it gives, but not the colour or an image of the colour. Although it might be true that someone who has the concept might not be able to use it, because he is blind, this fact needs no explanation and calls for no special theory. Indeed, equipped with the right definitions and suitable colorimetric equipment it might enable a blind man to pick out the colour. (Does the dictionary entry for 'uncle' enable us to pick out uncles? Would it enable a blind man to identify uncles on the street?) But there is surely something the words will not do. They will not immediately *acquaint* the unacquainted inquirer with the definiendum. Well, if *this* is what has to happen before a word can be unpacked, then it is very doubtful whether any word can be unpacked. 'Male sibling of a parent' does not acquaint us with any uncles.

Locke says:

> For, words being sounds can produce in us no other simple ideas than of those sounds; nor excite any in us, but by that voluntary connexion which is known to be between them and

> those simple ideas which common use has made them signs of. He that thinks otherwise, let him try if any words can give him the taste of a pine apple, and make him have the true idea of the relish of that celebrated delicious fruit.[9]

There are many troubles in this passage, most obviously the hopeless 'words being sounds' and the famous ambiguities of 'idea'.[10] The important point for us, however, is that we cannot define the taste of pineapple, because any definition, being conveyed to the *ear*, cannot give us a *taste*. But *no* definition could pass this test. The definition of an uncle will not give us one, if we lack it; the definition cannot produce a taste, but nor can it produce an uncle. We must distinguish the image, sensation or 'relish' from the concept or notion or definition.

Had Locke not worked with the ambiguous 'idea' idea he would have seen that though the words in any definition of the taste of pineapple (Locke's '*a* pine apple') cannot give us the idea or the sensation, this is equally true of the words used to define 'uncle' or 'triangle'. On the other hand, the idea in the sense of concept or notion of a triangle can be given by words, but, it seems, there is no such definition available for the taste of a pineapple. In this sense 'the taste of pineapple' cannot be unpacked. But I want to suggest that the comparison between 'taste of pineapple' and e.g. 'uncle' is an unfair one. The taste of pineapple is a *particular* taste, whereas 'uncle' does not stand for a particular uncle. The fair comparison would be between 'taste' and 'uncle' or between 'the taste of pineapple', a particular taste, and 'my uncle Bill', a particular uncle. The point is that Uncle Bill's 'relish' or phenomenological flavour is just as resistant to definition as the relish of pineapple is.

At the deepest level the complaint against definitions of *qualia*, such as the one given above for 'brown', does seem to be that the dictionary does not acquaint us with the *qualia*. The concealed requirement is that the definition

[9] *Essay*, III, iv, 11.

[10] Detailed by Ryle in 'Locke on the Human Understanding', *Collected Papers*, I (London: Hutchinson, 1971).

should not only define the word, say 'brown', but also in some sense *be* that colour, or at any rate somehow conjure it up, or excite the sensation in us. Perhaps Webster should put in a picture of brown, as some dictionaries do of certain birds. Should he also put in a picture of an uncle, Tom Cobley or Sam?

Stout seems to have wanted his scrutiny to yield not only the sensation of red, but also the separate sensation of its being the particular red it is. The impression his remarks make is that 'any kind of scrutiny' (what kinds are there?) is *staring* very hard. Staring very hard at gold does not reveal its stereotypical properties or marks, thinking very hard of a certain number does not enable us to discover that it has the interesting property of being: the smallest integer which is the sum of two cubes in two different ways.

The definition of a triangle will not *give* us a triangle, nor the sensation of one. But it will enable us to construct one out of given geometrical elements. In the same way one could place upon the expansion of a colour concept the requirement that it should enable us to construct or specify that colour by given elements and operations. This is in fact what is done in colorimetry, the theory of colour measurement.

A very important incidental point needs to be made. It cannot be a weakness in the geometrical definition of 'triangle' that the concepts which appear in the definiens and the definiendum are alike geometrical concepts. Nor is it a weakness in the definition of 'uncle' that the simpler concepts into which it is resolved are themselves socio-biological or kinship concepts of the same kind. And nor does it matter that 'scarlet' is defined by 'red' and 'yellow', or a taste by 'sweet', 'sour', etc.

The natural objection is that someone can be in possession of the definition of brown, or know the many marks of red (that it is advancing, hot, chromatic, a relatively dark colour at maximum saturation, mixes subtractively with white to make pink, etc.) without knowing what brown and red *are*. But in what sense does someone who does see brown know what brown is?

We know what the colour is only in the sense that we can

pick it out and distinguish it from other colours. We know it only in the Wittgensteinian sense that 'I have learnt English' is a proper answer to the question how I know that this colour is red.[11] In the same way, Johnny can know that this is an uncle, his uncle, call him affectionately 'uncle', without having a distinct idea what an uncle is. Johnny knows his uncle as such clearly but not distinctly. 'He knows what red is' can mean clear knowledge ('He knows what colour red is') or distinct knowledge ('He knows what the colour red is').

The question is whether colours can be known distinctly, or whether the empiricists are right and there is nothing to know here. If the suggested definition of brown is rejected because it fails to capture the qualitative essence of the quality, then colours will be simple, but so will everything else, and no analytical definition will succeed. Perhaps the definition does not capture the full phenomenal quiddity of brown; nor is it intended to. Nor does the definition of 'uncle' capture the full phenomenal quiddity of uncles, nor does 'figure obtained by joining in pairs three non-collinear points in a plane by straight line segments' capture the full phenomenal quiddity of triangles. ('That's not a triangle, that's a lot of words.') Definitions do not *give* things, they are *of* things. What they give is concepts. And if the proposed marks are attacked in *this* way, all definitions will go with them, for no words can capture the quiddity, phenomenal or otherwise, of anything. So much the worse for quiddity. What has clearly gone wrong is not that brown in its full brownness resists the net of concepts, but that a fantastically high standard has been set for what the concepts are expected to achieve. And then it is complained, and the complaint elevated into a philosophical or phenomenological discovery, that this standard is never met.

Does the definiens for 'brown' leave out the *brownness* of brown? *Shouldn't* definitions leave out what they define? If they didn't they would be circular! We must inquire what

[11] *Philosophical Investigations*, 381.

the brownness of brown, the redness of red, and the yellowness of yellow, etc., are, and how these differ from the colours brown, red and yellow.

The 'ness' suffix is typically used to indicate a state or condition, or the significant and remarkable degree to which something is in that state or condition. Thus the greenness of the tree is the striking degree to which it is green. The redness of his face, which indicates his guilt, is not the colour red but the fact and the suggestive extent to which his face is coloured red. The blueness of the winter sky is the fact that it is blue and the glorious extent to which it is – a degree. But what is the greenness of *green*? Is green unusual among the colours in being so strikingly green? Are the other colours then less green than green? The truth is that the 'is' in 'green is green' is the 'is' of identity, whereas the 'is' in 'The tree is green' is the 'is' of predication; the tree is green coloured, but it is not the same thing as the colour green. The greenness of the tree is the fact that it is green, not something in addition to the colour. The greenness of green is then the degenerate fact that it is green. Perhaps the metaphysical misuse of the suffix takes off from the fact that it is also used to indicate uncertainty or ignorance or to reduce commitment about what it is that is coloured green. A greenness in the sky may be an unidentified flying object, but if we aren't sure we may not want to commit ourselves. 'There was a greenness in the sky' – a being green, but of what we don't know. So 'greenness' naturally comes to mean the colour as it were detached from the coloured object. But 'greenness' is not the name of a colour. In certain quasi-mystical moods the greenness of green can come to seem remarkable – but is there any option? The *pinkness* of green? The yellowness of green? (but cf. the yellowness of *a* green). In *Metaphysics* Z[12] Aristotle says quite rightly that 'to ask why a thing is itself is no question at all [a meaningless inquiry]'. He adds that we can ask why a man is an animal of such-and-such a kind, and in the same way we can ask why some object is such-and-such a colour, in the sense of being coloured it. I

[12] 17, 1041^{a}15.

would be the last to defend physicalism, but it cannot be a proper objection to the physicalist theory of colours that 'red is a light emission of approximately 650 nm.' leaves out the redness of red. Or rather, it does; but since the redness of red reduces to the colour red via the sameness of the being red of red and the fact that it is red, the proper form the criticism should take is that it is not a definition of the colour at all, something which is manifestly not a light emission of any wavelength. The definition is too broad, too narrow, and it does not state the essence. We should not fear that the right definition of red will miss the redness of red, the ineffable *this*ness of red. There is an interesting analogy between the conception of a colour in empiricist epistemology and the medieval notion of a *haecceitas*. For empiricism colour is *this* (something we can say not what) so *a* colour is *a* this. Of the absurd idea of *haecceitas* (to be sharply distinguished from the common-sense idea of essence) David Wiggins has argued that its semantic impropriety and unlovely sound are fully matched by its logical incoherence. By attempting to perform both the linguistic functions of designating and predicating it can succeed at neither.[13] 'This is red', as we have seen, is ambiguous. It can mean either 'This colour is the colour red', which is a designation, or 'This object is coloured red', which is a predication; but not both. The confused empiricist notion of a colour-patch, which is supposed to be the ontological correlate of 'this', is an attempt to have it both ways. Is a colour-patch a colour or a patch? It can hardly be both. If it is a colour 'This is red' is a designation. If it is a coloured patch or a patch of something coloured, 'This is red' is a predication. The colour-patch, a logically particularized manifestation, will like a *haecceitas* always defy description, because it is a confusion, a kind of metaphysical shadow thrown up by the radical misunderstanding of 'this'. To be *this* is not to possess an extra essence, but just to be whatever I happen to be pointing at or drawing attention to. (And if there were such a thing as a *haecceitas*, why not also an *illuditas*, or for that matter the

[13] David Wiggins, *Substance and Sameness* (Oxford: Blackwell, 1980), p. 120.

practically existentialist *aliuditas* – thisness, thatness, and the otherness?) The general point that needs to be made is that 'this' is not a substantive and 'colour-patch' is not English (cf. 'patch-colour'). What is a 'red-patch' if not a patch of something red or a red coloured patch? The hyphen is worth noticing. A colour-patch as it were hovers between the two linguistic functions. It prevents the colour or the patch being definitely identified as the subject. But then the predicate collapses. A similar purpose is served by the omission of the verb in that famous non-sentence 'Red-patch now', which as it stands hovers between the alternatives and doesn't mean anything at all. If we feel that we must finally say with the scholastics that *individuum est ineffabile* it must be our understanding of the nature of the individual (colour, in this case) which falls under suspicion.

We have strayed rather far from the main theme, and it is now time to return to it. We left Pears saying that colours 'contain no separately noticeable marks'. Presumably what he means is Aristotelian, not Platonic, that one mark of tomatoes (their distinctive bulgy shape) can be noticed separately from the tomatoes, not separately from anything at all, and that there is nothing in a particular colour which can be noticed in any other colour. But this isn't true. The brightness of yellow, for example, which is a very important mark of yellow, can be noticed separately, in creams and whites. The unsaturatedness of a brown shade can be noticed in other shades of colour which have the same degree of unsaturatedness. Pears does not mean that the marks are not somehow visually distinct in the colour before us, like the features of a face – or does he? What does it mean for something to *contain* a mark, and why is this word chosen? It is of course a metaphor, but a very misleading spatial one. Whatever has marks must have them, not in any literal sense contain them. Would a colour have to contain marks in the sense in which a beer can contains beer, or the USA contains Kansas, or a tartan contains stripes (or is it made of stripes?) or in which a particular fabric pattern contains spots? What Pears is depending on is the visual impact of the spatial uniformity of a coloured area. It is the uniformity of the area which tempts us to think that the colour lacks marks. The marks of the colour are confused with marks on the coloured area. But these are marks in a different sense – discolorations. What would it be for a colour to contain marks in this literal sense? Consider a red patch surrounded by an orange patch. Is this what it means to say that orange contains red – a mark of orange? The meaning is rather that orange is to be regarded as a mixture of red and yellow; and in a certain sense (the colour printer's sense) the red in the orange can be separated. Red and yellow are the *separations* for orange, and in my view they can be 'noticed' in orange. Orange is *noticeably* yellowish. But the orange is a genuine mixture, such that no point on the orange patch is colour-distinguishable from any other. Stout is really requiring that a mixture should *not* be a mixture. The distinguishable and separable constituents of orange are red and yellow, but an orange surface is not a red surface *and* a yellow surface, or a surface of yellow and red spots – except when these are viewed at a distance. Can one discern the soda in a whisky and soda separately? It doesn't taste like neat soda. But the soda is noticeable. Can the taste of whisky and soda be scrutinized and the complexity detected? The central confusion here seems to be between the logical simplicity of what is experienced (a taste) and what psychologists have called the 'unitariness' of the experience.

The notion of a mark is a very interesting one, quite apart from its Leibnizian and Fregean uses. Partridge's *Origins* gives OE *merc*, a limit or boundary, hence boundary sign, limit, goal, target or indication thereof. It is as if Pears had confused the boundary with the visible boundary *sign* (the mark of the mark). Finding no marks in the second sense, no *noticeable* marks, he concludes that there are none in the first. The *OED* gives under 'mark' 'to separate *as by* a line' not 'by a line'; the line need not be drawn in ink or marked by a fence. The colours in the colour circle are separated as by a line, but the line need not be drawn in.

What kinds of marks has Pears failed to find? Does a colour have a mark like a hallmark or a watermark, a birthmark or a mastermark on a weight? Clearly not. But isn't this what 'separately noticeable' asks for? Certainly red is marked off by 'natural' boundaries from the other colours in the colour circle. But it doesn't *contain* these boundaries, nor are they visible by any kind of scrutiny. Consider Fig. IX, illustrating the boundaries of salmon, a so-called triplex colour whose appearance reflects the 60 per cent white, 20 per cent orange and 20 per cent yellow balance of the pigments used to mix it. These proportions give the boundaries or marks of salmon, but they are not to be found by visually scrutinizing the coloured area.

I think that what has happened is that a false and crude image of what a mark must be has been allowed to set itself up, the kind of thing appropriate to a piece of pottery or old silver or a banknote, and then of course no such thing can be found by any kind of scrutiny. In yet another sense marks are discolorations, spots. Here also a false imagery can suggest itself. Colours, such as the colour red, cannot be discoloured, though coloured patches can be. So colours have no marks. So they are simple.

Pears's view depends on something like what Geach has called 'a false Platonistic logic': 'an attribute is thought of as an identifiable object'.[14] This is part of the general doctrine Geach calls *abstractionism*, the theory that 'a concept is acquired by a process of singling out in attention some one feature given in direct experience – *abstracting* it – and ignoring other features simultaneously given – *abstracting* from them'.[15] It would be better, Geach argues,

> to follow Frege's example and compare attributes to mathematical functions. 'The square of' and 'the double of' stand for distinct functions, and in general the square of x is distinct from the double of x; but if x is 2 then the square of x *is* the double of x, and the two are not in this case distinguishable even in thought. Similarly, since what is chromatically coloured is often not red, 'the chromatic colour of x' or 'x's being chromatically coloured' does not in general stand for the same thing as 'the redness of x'; but if x is red, then it is by one and the same feature of x that x is made red

[14] P.T. Geach, *Mental Acts* (London: Routledge and Kegan Paul, 1971), p. 39.
[15] Ibid., 18.

Fig. IX

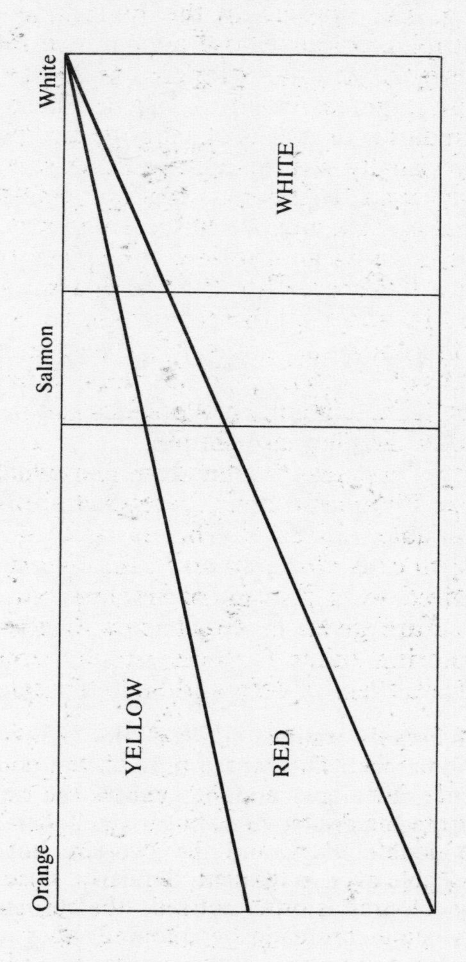

PROPORTIONS OF COLOURED PIGMENTS NEEDED TO MIX WHITES, SALMONS AND ORANGES

> and chromatically coloured; the two functions, so to speak, assume the same value for the argument x.[16]

I agree with Geach that abstractionism is as bankrupt as he says it is. Where I differ from him is in thinking that we should feel inclined 'to say that the mind makes a distinction between redness and chromatic colour, although there are not two distinct features to be found in the red glass or in my visual sensation'.[17] The question I have for him is exactly how a 'distinct feature' would have to be conceived, and how this conception could intelligibly apply to 'chromatic' or 'saturated'. Would the features have to be *visually* distinct or *spatially* distinct, like the features of a face, in order to count as distinct features? What does 'distinct ... in the red glass' actually mean? The two distinct features that Geach says are not here in the glass are, it seems to me, misconceived at the outset, and I think that he has himself accepted the key assumption of the false Platonistic logic. What *is* distinctness if the saturation of a colour is not distinct from the colour itself? Other colours can have the same degree of saturation (and saturation is essentially a degree concept) and in this sense saturation is distinct from colour and chromaticity, just as a line is distinct from the length of a line, a moving object from the velocity of the object, the acceleration from the velocity, and the volume and pitch of a sound from the timbre. But it does not follow that pitch, volume and timbre can exist independently from one another. They are distinct but interdependent dimensions of sound, in the sense that there cannot be a volume all by itself with no pitch or timbre. I am with Thomas Reid here, in his lucid comments on Berkeley's principle that we cannot 'abstract from one another, or conceive separately, these qualities which it is impossible should exist so separated'. Of Hume, Reid says,

> The precise length of a line, he says, is not distinguishable from the line. When I say, *this is a line*, I say and mean one

[16] Ibid., 39.
[17] Ibid., 38.

> thing. When I say *it is a line of three inches*, I say and mean another thing. If this be not to distinguish the precise length from the line, I know not what it is to distinguish.[18]

I have suggested that the origin of the imagery which lies behind the dogma that colours are simple is in the confusion between a colour and a coloured patch. This confusion is at the bottom of empiricist epistemology, and this is very clearly brought out by Lilly-Marlene Russow in her article on 'Simple Ideas and Resemblances' in Hume.[19] She points out that for Hume the terms 'different', 'distinguishable' and 'separable' are coextensive. 'Simple perceptions or impressions and ideas are such as admit of no distinction or separation.'[20] This means (Russow's example) that we cannot speak of the 'tonal quality' of a sound and the volume as 'two different qualities' because they cannot be separated. We cannot have the tonal quality without the sound. I am not sure what a tonal quality is, but suppose for the purposes of argument that it is pitch. Now surely we can diminish the volume of a sound without altering the pitch or timbre, otherwise no sound could grow fainter or fade away. (The correct understanding of such persistence through change is *not* that one sound slice is replaced immediately by another just noticeably different sound slice of slightly lower volume.) So the pitch can be separated from the volume. But it cannot be separated from volume – although the converse is not true. Bass drums have no pitch. It is not clear whether Hume's 'separable' means that the volume must be separable from the particular and determinate pitch manifestation for the sound as a whole to count as complex, or whether it must be separable from the determinable *pitch* – any pitch. His conclusion may depend on an equivocation between these two.

The white globe, or rather the idea of it, is said to be simple because we cannot 'separate and distinguish the

[18] Thomas Reid, *Essays on the Intellectual Powers of Man* (Cambridge, Mass.: MIT, 1969), p. 518.
[19] *The Philosophical Quarterly* **30**, No. 121 (1980), p. 342.
[20] Russow, 'Simple Ideas and Resemblances', p. 345. Note that it is perceptions of colour that are said to be simple here, not colours.

colour from the form'. But of course we can – though not with a pair of tweezers. Hume's white marble cube provides the perfect example. We can have the idea of something which is the very same colour as the white globe which is not the white globe. So either the globe is inseparable from the cube, or the colour is not inseparable from the globe. As for the determinable we can, I think, have the idea of something white without having the idea that it has any shape at all. We can have the idea (= sensation) of something flashing by too quickly for us to see or determine its shape, a white blur. Although it must no doubt have some shape, our idea or sensation may be indeterminate in this respect. And we can also have or understand the idea (= notion, concept) of something of whose shape we have no idea. I can have the idea that there is something white on my desk, without having any idea what shape it is – 'I know I saw something white on my desk'. It is just completely false that 'A person, who desires to consider the figure of a globe without thinking on its colour, desires an impossibility'[21] or else geometry and colour science could not be practised separately. 'Imagine a globe with a radius of three inches' – 'I can't, until you tell me what colour it is'. And 'Think of a colour' does not mean, 'Think of a colour which is the colour of something which has a shape that you have in mind', nor need there be any shape we think of when we think of a colour. Hume does say, it is true, that if we compare a white and a black globe 'we find two separate resemblances in what formerly seemed, and really is, perfectly inseparable'.[22] This is very similar to Geach's suggestion that it is the very same feature of x that makes it red and chromatically coloured, though the two functions are different ones. We rejected this proposal in Geach's logical formulation. Hume's psychological formulation has difficulties of its own. After practising, he says, we learn to distinguish shape and colour by a *distinction of reason*. We consider shape and colour together because

[21] Hume, *A Treatise of Human Nature*, I, i, 7.
[22] Russow, 'Simple Ideas and Resemblances', p. 345.

> they are, in effect, the same and undistinguishable; but [we] still view them in different aspects, according to the resemblances of which they are susceptible. When we would consider only the figure of the globe of white marble, we form in reality an idea both of the figure and the colour, but tacitly carry our eye to its resemblance with the globe of black marble: and in the same manner, when we would consider its colour only, we turn our view to its resemblance with the cube of white marble.[23]

This is plainly unsatisfactory. Why are things susceptible to the resemblances they are, if not because they really do have distinct and different aspects by which they are distinguished and compared? How can colour and shape be the same and indistinguishable, and yet have utterly different relations? What does it really come down to to 'view them in different aspects'? If we form the idea of a triangle, are we really holding in mind two triangular bits of coloured paper, and ignoring the colour? And how does this differ exactly from having the idea of a triangle all by itself, as it were in a separate compartment of the mind?

About the theory of simple ideas generally, Russow makes just the right observation:

> Hume can consistently claim that the idea of a white globe is simple, and the idea of an apple is complex. There is a problem here, but only a minor one: in the light of his remarks about the white globe, it seems to follow that the colour-of-the-apple-plus-the-shape-of-the-apple is a simple idea, but that the colour alone, being inseparable from the shape, is not. Similarly the idea of a shade of blue would be more accurately described as the idea of a blue patch.[24]

'Hume's general tendency to talk about colours *per se* as simple is not consistent with his considered analysis', Russow adds. The point is not minor for us. What is simple in Hume's account is not the colour blue but the blue-patch, the logically particularized manifestation discussed earlier. The confusions between logical simplicity, visual uniformity

[23] Hume, *A Treatise of Human Nature*, I, i, 7.
[24] Russow, 'Simple Ideas and Resemblances', p. 346.

and also psychological unitariness, all as it were housed in the one phenomenal atom, are made inevitable if we start with coloured areas as our data, or with the multiply ambiguous psychological/logical 'idea'. These confusions also attack the notion of a mark. A mark can either be a kind of spot, something which prevents visual uniformity, or a logical characteristic which is not necessarily phenomenal. If we return to Locke's original account of the simplicity of simple ideas, we find the confusions actually present in the definition, as well as a clear indication of the main ambiguity in his use of 'idea'. The simple ideas, 'being each in itself uncompounded, contains in it nothing but *one uniform appearance, or conception in the mind*, and is not distinguishable into different ideas'.[25] Here Locke is actually defining simplicity as uniformity of appearance. The dogma was a confused one from the start, depending as it did on the spatialization of colour. In the next chapter I show how this spatialization of colour has its origins in a particular theory of colour and a particular mode of scientific theorizing.

[25] Locke, Essay, II, ii, 1.

9

Sensations and Science

Seht ihr den Mond dort stehen.
Es ist nur halb zu sehen.
Und ist doch rund und schön.

MATTHIAS CLAUDIUS, *Abendlied*

9.1 Colours as Visual Units: The Mosaic Conception

If colours are spatialized and identified with patches of colour in the visual field, they become homogeneous elements, each of which could be lifted out of the array and replaced by another. Each colour-patch is identical with every other of the same colour, and differs *solo numero*. Each colour-patch differs from colour-patches of other colours in just one indefinable respect. Thus we arrive at a conception of the contents of visual experience in which it is composed, like a mosaic, of uniform interchangeable *pieces* of colour. It is from this idea – the Mosaic Conception, as I propose to call it – that the puzzle propositions represent a specific problem. Why, after all, *shouldn't* white be the lightest colour? Doesn't some colour have to be? Or do we think there is a uniformity here? The elements of the Mosaic Conception are uniform elements of fixed brightness which are simple and logically distinct from one another. Each element is spatial, and together they form a spatial array. Yet the puzzle propositions suggest that the uniform elements of the Mosaic cannot be either simple or logically distinct. It would seem to be a defining property of white that it must be lighter than any other colour. Describing an inclination which he himself felt, Wittgenstein wrote,

> We are inclined to believe the analysis of our colour concepts would lead ultimately to the *colours of places* in our visual field, which are independent of any spatial or physical

> interpretation; for here there is neither light nor shadow, nor high-light, etc. etc.[1]

This is a refined version of the Mosaic Conception, or, one might say, the Mosaic in its first stage of crystallization. (Note by the way that in Wittgenstein's remarks, what *we* are inclined to say must be resisted, what *I* am inclined to say is offered as the truth.) Wittgenstein does not introduce into his characterization the extra mistake of turning places in the visual field into literal *parts* of the visual field. Colour and place are however confused in the sense of Helen Knight. Colour is also independent of spatial and physical interpretation, however, in the different sense that, by themselves, colour places do not contain information concerning texture, grain, shadow, edge, highlight, transparency and so on. So there is also nothing in the colour place which would identify anything in the visual field as part of a material object.

In the Mosaic Conception the analysis of colours serves an epistemological purpose. Colour concepts are to be built up out of the uniform elements of the visual field. It is from the point of view of these epistemological purposes that the attribution of colours to things represents the philosophical problem that it does, e.g. the fact that white bread is not white, or that a grey element turns out to be of something white in shade when the physical interpretation is restored, or that blue stars and diamonds are not really blue, etc. In much of the *Remarks on Colour* Wittgenstein is struggling to free himself from the Mosaic Conception and its unresolved residue in our patterns of thinking about colours. Yet the epistemological purposes served by the Conception are in a sense very limited. We can imagine the epistemological procedure of constructing it on the model of a game. The object of the game is to fill in places in the visual field, as in a child's 'colour by number' book, until a complete 'picture' is built up. The game consists of calling out possible names of colours of patches of various sizes. If a patch consists of more than one colour, the response is 'No', and

[1] Remarks on Colour, I 61.

the game is continued with a smaller patch. If a patch is an absolutely determinate shade, a colour word can be accepted, and the game continues with another patch. The game ends when the entire picture is, in a misleading idiom, filled with colours. The ultimate simple elements reached, however, are simple elements only in the sense in which a square on a chessboard could be regarded as simple or indivisible. No square can be occupied by more than one piece. But this says nothing about simplicity in any physical or absolute context.

9.2 Simplicity Again

The conception of colours for which I have been arguing goes against the Mosaic Conception, precisely in that it makes colours essentially *not* part of a visual field. On the contrary, the puzzle questions require precisely the metaphysical perspective by means of which Leibniz explicates the meaning or 'signification' of natural kind terms, with the difference that 'inner constitution' is not meant literally. For a colour is not a body.

> So you see, sir, that the name 'gold', for instance, signifies not merely what the speaker knows of gold, e.g. something yellow and very heavy, but also that which he does not know, but which may be known about gold by someone else, namely, a body endowed with an inner constitution from which flow its colour and weight, and which also generates the other properties which he acknowledges to be better known by the experts.[2]

If it was a necessary truth that

> When we're asked, 'What do the words "red", "blue", "black", "white", mean?' we can of course immediately point to things which have these colours, – but our ability to explain the meaning of these words goes no further! For the rest, we have either no idea at all of their use, or a very rough and to some extent false one.[3]

[2] G.W. Leibniz, *New Essays on Human Understanding*, trans. and ed. Peter Remnant and Jonathan Bennett (Cambridge: Cambridge University Press, 1981), III xi 24, p. 354.

[3] *Remarks on Colour*, I 68.

then the puzzle propositions would be unanalysable and the puzzle questions unanswerable. Since the puzzle questions can be answered, our ability to 'explain the meanings' of the colour words *does* go further. The question then will become what form this explanation must take. Since there must be a theory of meaning which gives a true account of the meaning of colour words, then, given that colour words have the type of meaning that they do, we ought to ask what form this theory must take. Those who wish to deny that there *could* be anything which would count as 'our ability to explain further' must do much more than causes us to feel the urgency of the idea that no verbal technique, being *verbal*, could capture the essence of the uniform and featureless white space before us. *Remarks on Colour* I 68 is followed by the suggestive I 69: 'I can imagine a logician who tells us that he has now succeeded in *really* being able to *think* 2 × 2 = 4.' Perhaps this is intended as a reproof for Russell. Yet if a logician did succeed in doing something which deserved to be called 'being able to think 2 × 2 = 4', say by Russell's method, why should such a technique be rejected merely because its form happened not to coincide with that of the sequence of half-memorized mumblings which most of us go through when confronted with a grocery bill? Possession of the ordinary concept of a number is in general little more than an ability to manipulate a symbol in standard sequences and procedures. Similarly, the ordinary concept of a colour, or the use of an English colour word, can enable us to tell others *which* colour we have seen, but not *what* it is that we have seen. The fact that knowledge is clear in Leibniz's sense does not mean it cannot be made distinct.[4] Even though 'white' obviously *does* mean *this* – white – colour words need not be treated as meaning *this* or *this* or *this*, as they are in the empiricist epistemology of the Mosaic Conception. Here I follow the anti-abstractionist *Remarks on Colour*, III 25: 'Why isn't a saturated colour simply: *this*, or *this*, or *this*, or *this*? – Because we recognize or determine it in a different way.'

[4] 'Meditations on Knowledge, Truth and Ideas' in *Philosophical Papers and Letters*, ed. Leroy E. Loemker (Dordrecht: Reidel, 1976), p. 291.

In the previous chapters I have tried to show how the explication of the puzzle propositions demands that light itself and shadow, or darkening, enter into the colours. If colours were in the end essentially the contents of places in the visual field, then there would be, as Wittgenstein says, no light and shade in them. Thus the confusion of colours and place in Helen Knight's sense leads to the denial of the implication of light and shade in the colours. This would make the proposed real definitions impossible. 'Light' in the visual field would be strictly adjectival, meaning not the stuff light, but a degree of brightness of a place. Here the phenomenal becomes the adjectival, and the substantive is the physical. The phenomenal is also identified with the mental, so that the mental becomes the adjectival. Since complexity requires the substantive, *the mental becomes the simple*.

9.3 Origins of the Mosaic Conception in Science

Though it appears to be powered by purely epistemological considerations, I believe that the Mosaic Conception has its real historical origins in the impact of science on the world conception. Roughly, the idea is that science pushed the external world of visual facts – such as the blue of a certain light – into the mind of the perceiver where it became the Mosaic. When physics took over the description of the external world in the sixteenth and seventeenth century, it needed to divide the phenomena into those which would be handled by the emerging mathematical physics and those which would not. The external world shrank around the subject, and with the twin demands of scientific explanation and epistemological analysis, hardened into the Mosaic Conception.

In colour theory itself, the diremption of the data into physics, on the one hand, and the psychology of simple subjective sensations on the other, derives from a theory of the visual process which I shall call the Newton Model. Subjective sensations are counted simple if they consist of just one colour. This is a requirement of the Newton Model because each type of colour sensation must be placed at the end of a chain of events which constitute just one sort of physical explanation.

Newton divides the visual process into three distinct temporal parts. The first is physical, and understanding it is the responsibility of physics. Colours in the object perceived are 'nothing but the disposition' of that object to reflect a certain sort of Ray more than any other. The second part is physiological. The Rays propagate a motion in the sensorium. This is followed by a visual sensation of the physiological motions 'under the Forms of Colours'.[5]

9.4 The Newton Model and the Mind-Body Problem

The Newton Model has exerted an enormous influence on colour science since the seventeenth century. Yet it remains profoundly problematic. Consider the following philosophical exposition of the form of the Model from John Stuart Mill, which links the autonomy of colour as a phenomenon or sensation to states of consciousness and to the limits of scientific explanation.[6] Yet it turns on an essentially *trivial* point concerning the inexplicability or simplicity of colours.

> The ultimate laws of Nature cannot possibly be less numerous than the distinguishable sensations or other feelings of our nature ... Since there is a phenomenon sui generis, called colour, which our consciousness testifies to be not a particular degree of some other phenomenon ... but intrinsically unlike all others, it follows that there are ultimate laws of colour; that, although the facts of colour may admit explanation, they can never be explained from laws of heat or motion or odour alone, but that, however far the explanation may be carried, there will always remain in it a law of colour ... However diligent might be our scrutiny of the phenomena, whatever hidden links we might detect in the chain of causation terminating in the colour, the last link would still be a law of colour, not a law of motion nor of any

[5] Newton, *Opticks*, Prop. II, Theorem II (New York: Dover, 1959), pp. 124–125, and cf. W.D. Wright, 'Towards a Philosophy of Colour', in *The Rays Are Not Coloured* (London: Adam Hilger, 1967), pp. 19 ff.

[6] John Stuart Mill, *A System of Logic*, Book 3, Ch. 14, 2.

> other phenomenon whatever . . . The ideal limit, therefore, of the explanation of natural phenomena would be to show that each distinguishable variety of our sensations, or other states of consciousness, has only one sort of cause; that, for example, whenever we perceive a white colour, there is some one condition or set of conditions which is always present, and the presence of which always produces that sensation.

One might be favourably disposed towards this argument because it has an anti-reductionist conclusion: colour is not anything other than colour. This is of course true. The trouble is that in Mill's version of the Newton Model this is so, not because colour is not anything else, but because it is effectively nothing at all. I regard Mill's argument as a reductio ad absurdum of the Newton Model. Similarly, the manner in which the identity theory cuts the 'world-knot' – by *fiat* – is a formal recognition of the absurdity generated by the conception of perception implicit in the Newton Model.[7] The nomological danglers and the basic laws from which they dangle are problematic in the Newton Model because their role in it is essentially an idle one. Colours themselves play no functional physiological role, and they serve merely as the termini of the three-stage causal explanation. They could in fact play no role, for they are simples. Thus it achieves nothing of significance either to assert or to deny their existence at the third stage of the visual process. (All the really important mistakes have been made by this point.) Thus the Newton Model leads directly to an insoluble mind-body problem, insofar as it arises for the so-called phenomenal properties. Indeed, the last stage of the Newton Model *is* the mind-body problem. The specific sixteenth and seventeenth century scientific origins of the problem suggest that it is not after all a 'world-knot', but a knot in our own thinking. If, as some contemporary philosophers believe, the mind-body problem really is in-

[7] In a dialogue recorded with Urmson, Ryle observed that 'Australia is the last home of unconverted Cartesianism. Occasionalism is the same sort of doctrine as the empirical identity thesis, only I think the empirical identity thesis slightly better than occasionalism was, but occasionalism died quite soon after Descartes, and this should have died at the same time.'

soluble, we are entitled to put the question why *that* should be. The answer must inevitably concern the *form* of the problem. Its difficulty is due not to some elusive metaphysical fact or to an unobvious and small or easily overlooked logical point concerning, say, the relation between English statements about sensations and statements about brain processes, but *to the form of the problem situation faced by philosophers dealing with the effect of sixteenth and seventeenth century science. Physics* is what makes the mind-body problem really intractable. It is a very suggestive fact that Newton's original interest in the spectrum and the physics of dispersion was due to his attempt to solve the problem of chromatic aberration in refracting telescopes.[8] His problem situation was, how to *get rid of* the unwanted fringes of colour in the physical optical system. His account of the genesis of the spectral colours, in 'divers refrangibility' is in fact the image of his geometrical solution to *this* problem.

But the one thing which the Newton Model cannot explain, as Mill saw, is why a particular type of physical cause should be associated with a particular colour. The Newton Model cannot explain the content or qualitative character of sensations – *colours* – and nothing else could, because they are simple, beyond explanation. There *could* be no explanation of the fact that shortwave or small corpuscle light should finally cause violet-blue in the Mosaic rather than some other colour. Thus the one thing the Newton Model cannot explain is *colour*. This is not because of a metaphysical limit to explanation or because colour is identical with something which is not colour – some aspect of a brain-process – but *because of the structure of the Newton Model.*

Irvin Rock has given a sketch of explanation in the psychology of perception which illustrates the crucial infirmity of the Model.[9] Rock distinguishes three levels of explanation.

[8] Richard S. Westfall, *Never At Rest, A Biography of Isaac Newton* (Cambridge: Cambridge University Press, 1980), pp. 161–3; Louis Trenchard More, *Isaac Newton*, (New York: Dover, 1934), pp. 65–7.

[9] *An Introduction to Perception*, p. 6.

Objects and Events in the Real World

↓ energy or information coming to sense organs: light waves, sound waves, etc. (1)

Sense organs

↓ signals to brain (2)

Relevant Brain Events

↓ (3)

Perceptual Experience

According to Rock, when Newton discovered that sunlight is complex, and that the different components cause sensations of different hue, 'he thereby "explained" colour vision.' This truncated 'explanation' took place at Level (1). It was made complete with the identification of the 'proximal stimulus' in the retinal image. Explanations at Level (2) involve the description of the working of physiological mechanisms, e.g. that some receptors respond more to light of one wavelength than to light of another. This begins to make up for the deficiency of the Level (1) explanation, but it does not explain why shortwave light is seen in *blue* rather than some other colour. Why should we see blue when such and such a class of cone cell is stimulated by the shortwave light? What Rock calls a 'central explanation' is his answer. By this 'we mean an explanation in terms of the relevant events that occur in the brain that give rise to (or cause) the perception in question.'

This is a fairly common way of thinking in psychology, and the idea of *a perception* floating off, being given rise to, though clearly false, appears to be a very natural one, to judge by the number of introductory pyschology texts which are disfigured by illustrations depicting often just one sensation or perception 'arising' in a balloon above the subject's head. And of course Rock's 'central explanation' does *not* explain why the relevant events should cause the balloon to be filled with blue rather than some other colour. Note also that Level (1) Events and Objects take capitals and are a part of the Real World, a status by implication denied to Perceptual Experiences at Level (3). Then there is the question why the arrow signifying Level

(3) explanations has no name. Would 'Signals to the Mind' do? But the Mind would then have the task of 'interpreting' these signals and synthesizing them into perceptions, and the question would arise how – by what means or mechanisms – it could accomplish this. For here we are beyond all mechanisms. The contents of the Mind are simple and inexplicable. To add mystery to confusion, they are also held somehow to *constitute* consciousness itself.

9.5 Concepts of Consciousness

'What actually is the "world" of consciousness?', Wittgenstein asks at *Remarks on Colour*, III 316.

> That which is in my consciousness: what I am now seeing, hearing, feeling . . . – And what, for example, am I now seeing? The answer to that cannot be: 'Well, *all that*' accompanied by a sweeping gesture.

But of course this *is* virtually all there is to our notion of visual consciousness. It is significant that the words and the gesture which Wittgenstein rejects confuse consciousness and the external world. Wittgenstein's *question* confuses the substance of consciousness ('world') with the content of consciousness ('that'). In a preceding remark he put the question, 'What actually is the '*world*' of consciousness?'.

'In my consciousness', 'in my visual field' and equivalent phrases, most notoriously 'in the mind', typically allow two very different types of interpretation. In one what is in my consciousness is merely something – tower, mountain, bridge – in the external world which I happen to be attending to, and my consciousness has a sort of indexical function. Here consciousness, if we can put the matter this way, has no literal content. 'In my sights' does not mean 'Inside my gunsights'. In the second interpretation what is in my consciousness is literally inside my consciousness, in the sense that it composes consciousness. What is in my consciousness are 'contents', not external objects. These contents are in fact the appearances of real things, made into simple elements in the Mosaic of the visual field, and attached to their physical bases by a string of correlations.

9.6 Criticism of Sensations

The content concept of consciousness derives, as I have suggested, from the effect of physics on the thoughts of philosophers and scientists. The Newton Model forced colours into consciousness in a literal sense, where they were grafted into the diaphanous acts of consciousness by which, in the direct realist picture which the Newton Model replaced, they had been apprehended. The result was our present hybrid conception of a sensation, something which is somehow both act and object to itself. It is not that the act/object distinction for some interesting reason breaks down for sensations, but rather that in the notion act and object are confused. (In this sense the Newton Model is *idealist* about colours.) This confusion becomes especially apparent if we try to specify a principle by which sensations can be counted. The confusion surrounding the individuation of sensations and the conditions for their identity also derives, not from some significant feature of Mind, but from the fact that one principle cannot incorporate the requirements of both act and object individuation.

Consider the diagrams below. In A we have one sensation of white if we apply a content test, in B two. Here the

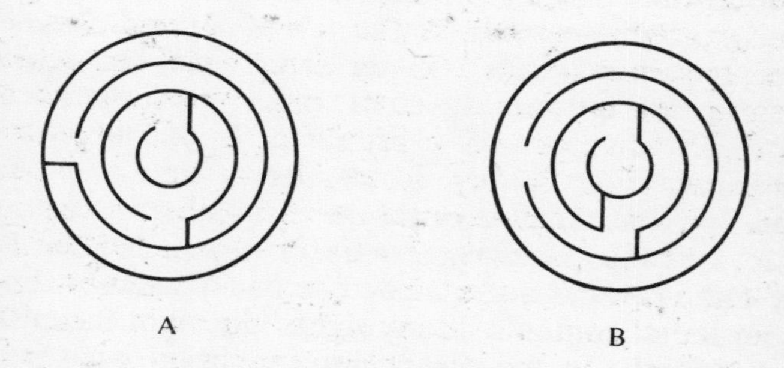

objects are unbroken areas of white. Yet I think we want to reject the result that we have two sensations of B and one in A. This is the pull of an act test. We are not aware of performing two acts when we look at B. Our awareness is not genuinely divided in B, as it is in C. If we were to make a careful study of B, we would have more than one sensation, but probably also more than two. Here as well we would be applying an act test.

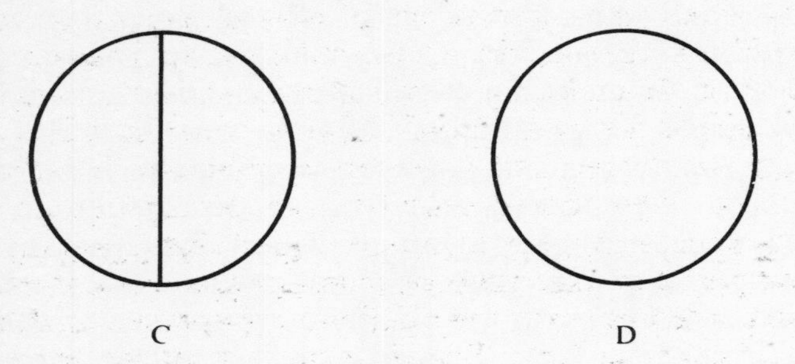

D certainly gives one sensation, whichever test we apply. So what makes D and A count as one sensation each, while B and C count as two? In respect of plurality are D and A really more alike than A and B? Are B and C more alike than C and D?

What makes A and B the source of such puzzling questions is that in them the act and content tests conflict. In B there is one act, and so, we think, one sensation. But there are certainly two objects or contents. So, perhaps, we begin to think that there is a need for a distinction between the content and the object, as well as between the sensation and the content and the having of the sensation. Perhaps also A and B exhibit a Gestalt which unifies the contents. Here a hazy pyschology begins to surface, and we are tempted to invent all sorts of logical or other distinctions – having, contents, objects, and so forth. It would be better to drop the whole flat-footed conception of a sensation, and look elsewhere for the basic receptivity of the subject in perception.

I would accept the term '(a) sensation' for such things as a choking sensation, a prickling sensation, and a gliding sensation, where what is sensed is a felt alteration in one's own bodily state rather than something in the external world, and also typically an alteration of an unusual kind or to an unusual or extreme degree. Thus I would also accept afterimages as sensations in good standing. So I should stress that nothing I have said so far bears upon the mind-body problem as it arises for sensations in this proper sense. But it is simply incorrect to generalize the idea and to say that when I look at a printed page I have any sensations of black and white, still less an indeterminate number of them. In my view sensations in either sense are not required for perception. Anyone who finds this paradoxical has not realized the extent to which he is in the grip of theory. Why should sensation, or perception, be composed of sensations? If someone argued that digestion must be composed of digestings or digestions, we would need to be persuaded of some very definite advantages to this way of looking at the facts. This does not, of course, mean that a Moorean direct realist theory is the right way of looking at anything.

9.7 Ryle on Perception

If the notion of *a* visual sensation is abandoned, what account can be given of the scientific knowledge gained under the auspices of the Newton Model? It is natural to want to separate the false metaphysics of sensation in the third psychological part of the Model from the physics and physiology of the more scientific part. A strategy of this sort is pursued by Ryle in his penetrating essay on 'Sensation',[10] and also in Chapter 7 of *The Concept of Mind*. There can be no quarrelling with the causal theory of perception, Ryle says, 'whether we are thinking of the stages covered by optics or acoustics, or whether we are thinking of the stages covered by physiology and neurophysiology. The final stage, covering a supposed jump from neural impulses in the body to mental experiences, or sense-impressions, is, however, quite a different matter'. Ryle says that we must distinguish causal questions about perception from questions about perception considered as an acquired skill. The second sort of question is a question 'about, so to speak, the *crafts* or *arts* of finding things out

[10] Gilbert Ryle, 'Sensation', in *Collected Papers* **2** (London: Hutchinson, 1971).

by seeing and hearing...' We are to understand perception or perceiving as a success or victory in the game of exploring the world, and not as a process of being happened to, with causal antecedents. To perceive, according to Ryle, is to be successful at something one is doing. But here it seems to me that Ryle conceded too much. Like Wittgenstein, he went wrong in thinking that science could only be an investigation into causes, as laid down for example in the Newton Model, so that there could not be a science of perception as he meant it. He restricted science to the investigation of causal antecedents, and then – rightly – denied the implications of causes in perception.

9.8 Wittgenstein, Gibson and Science

But why should there not be a study of *how* success is achieved in exploring the world? Why would such a study have to be an exclusively conceptual one? Would it have to be distinct from the understanding of the physiology of the senses, even if their function was not understood as the delivery of sensations? Must we believe that the only possible type of explanation in the psychology of perception is one in which e.g. the appearance of white is explained by showing how it 'arises' at the end of a hypothesized causal chain? I would take this in part as the question how sensation without sensations is to be understood, and I agree with Ryle that sensation is actually a kind of perception, and not the other way round. The registering of changes in the illumination caused by an object is a paradigm example of perception as information pick-up in Gibson's sense, and of course is a 'higher order invariant' or stimulus ratio as he means it. It is information in his sense rather than a sensory quality of the object.[11] So the view which develops out of the solution to the puzzle propositions leads on to the kind of ecological optics developed by Gibson, and already anticipated by Goethe for colour perception, especially in connection with the adaptive and homeostatic functioning of the senses, and the corresponding description of nature or

[11] For Gibson's denial that there are such things, 'Are There Sensory Qualities of Objects?' in *Synthese*, **19** (1968–9), pp. 408–409.

the environment at the appropriate level. I hope that the considerations put forward in this book in connection with what kinds of things colours must be if Wittgenstein's puzzle questions are to be answered will further an appreciation of the possibilities of this exciting new point of view. I believe that it is a view which might have been profoundly congenial to the later Wittgenstein, and that his most enduring contribution to philosophy will turn out to have been the image of the sciences uninfluenced by the wrong primitive pictures and exactly responsive to the different types of subject matter and facts which they must explain. In the psychology and physiology of perception problem and method need not pass one another by, as they are bound to do if the Newton Model is correct. In Gibson's ecological optics we have the beginnings, at least, of a science of the senses and their relation to the external world in which the relevant facts and phenomena appear at their own proper level. '"What an intelligent man knows is hard to know." Does Goethe's contempt for laboratory experiment and his exhortation to us to go out and learn from untrammelled nature have anything to do with the idea that a hypothesis (interpreted in the wrong way) already falsifies the truth? And is it connected with the way I am now thinking of starting my book – with a description of nature?'[12] When Wittgenstein wrote that 'Man has to awaken to wonder – and so perhaps do peoples. Science is a way of sending him to sleep again'[13] he can only have meant, or *should* only have meant, not science but the terrible scientism of the times. Yet he too was a victim of scientistic patterns of thinking. I have tried to suggest that he did not appreciate the full potential of scientific explanation in the domain of colour, and that he need not have sought the understanding of the puzzle propositions beyond science and the phenomena in a quasi-transcendental 'logic'

[12] Wittgenstein, *Culture and Value*, pp. 10e–11e. We know that Wittgenstein had the German text of the *Farbenlehre* before him as he wrote the *Remarks on Colour* III in Oxford in 1951, and must have noted the experiments described in it. What he should have written was, 'Goethe's contempt for a *certain kind of* laboratory experiment...'.

[13] *Culture and Value* (Oxford: Blackwell, 1980), p. 5e.

of colour concepts. Science need not eliminate the phenomena, nor can it possibly be, in its essence, inimical to life. Gibson wrote in the Introduction to *The Ecological Approach to Visual Perception*,[14]

> The great virtue of the headrest, the bite-board, the exposure device, the tachistoscope, the darkroom with its points of light, and the laboratory with its carefully drawn pictorial stimuli was that they made it possible to study vision *experimentally*. The only way to be sure an observer sees what he says he sees is to set up an experimental situation and check him out. Experimental verification can be trusted. These controls, however, made it seem as if snapshot vision and aperture vision were the whole of it, or at least the only vision that could be studied. But, on the contrary, natural vision can be studied experimentally. The experiments to be reported in Part III on perception involve the providing of optical information instead of the imposing of optical stimulation. It is not true that 'the laboratory can never be like life.' The laboratory *must* be like life!

9.9 If a Lion Could Talk . . .

If it seems unfair to accuse Wittgenstein of scientism, consider his amazing remarks about what would happen if 'one of you suddenly grew a lion's head and began to roar'.[15] This would be extraordinary, Wittgenstein says, and after recovering from our surprise, we would 'have the case scientifically investigated', and 'if it were not for hurting him I would have him vivisected'. But why? What could immediate vivisection be expected to achieve? And what primitive picture of the techniques of science impels these words? 'And where would the miracle have got to? For it is clear that when we look at it this way, everything miraculous has disappeared.' So, Wittgenstein concludes, the scientific excludes the miraculous. 'Everything miraculous has disappeared.' But of course it has. The lion has been *killed!* Its living behaviour can no longer be studied. Would it have been wrong to try to understand what it meant when

[14] (Boston: Houghton Mifflin, 1979), p. 3.
[15] 'Wittgenstein's Lectures on Ethics', *The Philosophical Review*, 74 (1965).

it roared, if anything? If the roaring was a speaking, shouldn't we have tried to understand? And must this be as impossible as xi of Part II of the *Philosophical Investigations* insists ('If a lion could talk, we could not understand him')? We could begin by investigating the creature's habitat, diet, its sociability and its psychology. As well as anatomy, there is ecology and there is ethology. And with *these* sciences it is not as clear that science excludes wonder. But then anatomy is wonderful as well. 'The truth is that the scientific way of looking at a fact is not the way to look at it as if it were a miracle.' This cannot be a metaphysical proposition about the ultimate limits of science, but only an observation about the psychology of scientistic or technocratic methods. In the *Remarks on Colour* Wittgenstein did consider a way of doing science which aimed to preserve the miraculous. But he rejected it.

> Goethe's theory of the constitution of the colours of the spectrum has not proved to be an unsatisfactory theory, rather it isn't a theory at all. Nothing can be predicted with it. It is, rather, a vague schematic outline of the sort we find in James' psychology. Nor is there any *experimentum crucis* which would decide for or against the theory.[16]

I have argued for the truth of Wittgenstein's next remark, that 'Someone who agrees with Goethe believes that Goethe correctly identified the *nature* of colour.' What I have not been able to see is the truth of his subsequent observation that 'nature here is not what results from experiments, but it lies in the concept of colour'.

[16] *Remarks on Colour*, I 170.

Index